Advanced Data Management: For SQL, NoSQL, Cloud and Distributed Databases

ADVANCED DATA MANAGEMENT: FOR SQL, NOSQL, CLOUD AND DISTRIBUTED DATABASES

Edited by:

Adele Kuzmiakova

ARCLER PRESS

www.arclerpress.com

Advanced Data Management: For SQL, NoSQL, Cloud and Distributed Databases

Adele Kuzmiakova

Arcler Press
224 Shoreacres Road
Burlington, ON L7L 2H2
Canada
www.arclerpress.com
Email: orders@arclereducation.com

ISBN: 978-1-77469-868-6 (Hardcover)

ABOUT THE EDITOR

Adele Kuzmiakova is a machine learning engineer working at the intersection of machine learning, computer vision, and natural language processing. Adele attended Cornell University in New York, United States for her undergraduate studies. She studied engineering with a focus on applied math. Some of the deep learning problems Adele worked on include predicting air quality from public webcams, developing a real-time human movement tracking, using 3D computer vision to create 3D avatars from selfies in order to bring online clothes shopping closer to reality, and creating visual stories and photobooks from photos on mobile devices. She is also passionate about exchanging ideas and inspiring other people and acted as a workshop organizer at Women in Data Science conference in Geneva, Switzerland in 2022 and 2023.

TABLE OF CONTENTS

LIST OF FIGURES

LIST OF TABLES

LIST OF ABBREVIATIONS

1NF	First Normal Form
2NF	Second Normal Form
2PC	Two-Phase Commit
3NF	Third Normal Form
AI	Artificial Intelligence
APIs	Application Programming Interfaces
AWS	Amazon Web Services
BI	Business Intelligence
CAP	Consistency, Accessibility, Partitioning
CCPA	California Consumer Privacy Act
CDC	Change Data Capture
CEP	Complex Event Processing
CIO	Chief Information Officer
COBIT	Control Objectives for Information and Related Technologies
CSPs	Cloud Service Providers
CTEs	Common Table Expressions
DBaaS	Database-as-a-Service
DBMS	Database Management System
DDL	Data Definition Language
DMBOK	Data Management Body of Knowledge
DML	Data Manipulation Language
DRPs	Distributed Reliability Protocols
EDWs	Enterprise Data Warehouses
ER	Entity-Relationship
ETL	Extract, Transform, Load
FAQs	Frequently Asked Questions
FK	Foreign Key

GCP	Google Cloud Platform
GDPR	General Data Protection Regulation
GIS	Geographic Information Systems
GPUs	Graphics Processing Units
GUI	Graphical User Interface
HDFS	Hadoop Distributed File System
HIPAA	Health Insurance Portability and Accountability Act
HPC	Performance Computing
IaaS	Infrastructure-as-a-Service
IoT	Internet of Things
KMS	Key Management Services
MFA	Multi-Factor Authentication
ML	Machine Learning
MPP	Massively Parallel Processing
NAS	Network-Attached Storage
OLTP	Online Transaction Processing
PaaS	Platform-as-a-Service
PII	Personally Identifiable Information
PK	Primary Key
RBAC	Roles-Based Access Control
RDBMS	Relational Database Management Systems
RDF	Resource Description Framework
ROWA	Read-Once/Write-All
RPCs	Remote Procedure Calls
SLAs	Service Level Agreements
SQL	Structured Query Language
SSDs	Solid-State Disks
SSIS	SQL Server Integration Service
SSL	Secure Sockets Layer
SSN	Social Security Number
STaaS	Storage Space as a Service
TCO	Total Cost of Ownership
TDE	Transparent Information Encryption
TLS	Transport Layer Security

PREFACE

In today's data-driven world, the effective management of vast amount of information has become a critical challenge for organizations across industries. With the advent of new technologies and the exponential growth of data, professionals in the field of data management must continually adapt to stay ahead. Advanced data management encompasses a range of techniques and strategies that enable organizations to efficiently store, retrieve, and analyze their data, unlocking valuable insights and driving informed decision-making.

The book is structured into eight distinct chapters, each delving into specific aspects of advanced data management. While the chapters can be read independently, they have been meticulously organized to ensure a seamless and interconnected exploration of the subject matter.

The book starts with an introduction to Advanced Data Management in Chapter 1, providing readers with a broad overview of the evolving data management landscape, the challenges it poses, and the opportunities it presents. This foundation sets the stage for Chapter 2, which delves into the fundamentals of data management, covering key concepts, principles, and best practices that form the bedrock of effective database design and administration. Chapter 3 focuses on the relational data model and relational database constraints, offering an in-depth exploration of the principles that underpin the relational database paradigm. This chapter examines the nuances of data modeling, normalization, integrity constraints, and the relational algebra, empowering readers to design and manage robust relational databases.

In Chapter 4, we turn our attention to the Extract, Transform, Load (ETL) process, a critical component of data integration and warehousing. Through the lens of SQL, this chapter explores the intricacies of data extraction, transformation, and loading, enabling readers to leverage SQL's power for efficient data processing. The landscape of data management extends beyond traditional relational databases, and Chapter 5 is dedicated to exploring NoSQL databases. This chapter provides an in-depth examination of the various types of NoSQL databases, such as document stores, key-value stores, columnar stores, and graph databases. By understanding the strengths and limitations of these non-relational databases, readers can make informed decisions when it comes to selecting the appropriate data storage solution.

Chapter 6 delves into the realm of cloud databases and big data, exploring the transformative impact of cloud computing on data management. This chapter investigates the benefits and challenges of utilizing cloud-based solutions for data storage, processing, and analytics. Additionally, it examines the convergence of big data technologies with cloud databases, enabling organizations to harness the power of large-scale data analytics.

The intricacies of managing data in distributed environments take center stage in Chapter 7. This chapter provides a comprehensive examination of distributed database systems, addressing topics such as data fragmentation, replication, concurrency control, and distributed query processing. By understanding the complexities inherent in distributed databases, readers can design and manage scalable and fault-tolerant systems. Finally, in Chapter 8, we explore the paradigm of Cloud Databases and Database-as-a-Service (DBaaS). This chapter sheds light on the benefits and challenges of utilizing cloud-based database services, examining various DBaaS offerings and their implications for data management. Furthermore, it explores strategies for migration, scalability, and security in the context of cloud databases.

With a writing style that emphasizes clarity and accessibility, this book endeavors to make the intricacies of advanced data management easily understandable for readers at all levels of expertise. Whether you are a seasoned practitioner or a newcomer to the field, the fundamental concepts, advanced techniques, and real-world examples presented in this book will empower you to navigate the ever-evolving world of data management.

CHAPTER

1

Introduction to Advanced Data Management

CONTENT

1.1. INTRODUCTION

Data management encompasses the systematic procedures involved in arranging, securing, preserving, and utilizing data to facilitate organizational activities, informed judgments, and long-term objectives. The process encompasses a wide range of methodologies, regulations, protocols, and tools with the objective of guaranteeing the accuracy, accessibility, consistency, and protection of data throughout its entire duration (Winkelmann et al., 2017). In contemporary times, effective management of data holds significant significance in light of the prevalence of digital technology. Small and large companies alike accumulate extensive quantities of data from diverse origins, including transactions, customer engagements, sensors, social media, and other sources. The data possesses significant value, and proficient management is crucial for accessing insights, enhancing operational efficiency, fostering innovation, and attaining a competitive advantage (Chen et al., 2021) (Figure 1.1).

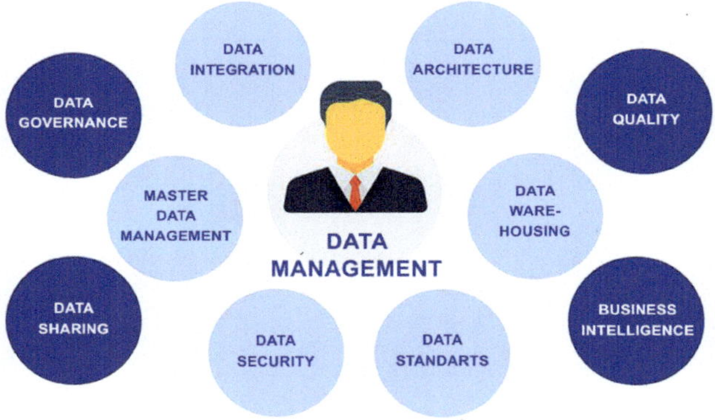

Figure 1.1. Illustration of data management.

Source: Data entry, Creative Commons License.

The implementation of rigorous data management practices can yield numerous advantages for businesses. The benefits encompassed in this context involve heightened precision and uniformity of data, augmented convenience and applicability of data, simplified amalgamation and dissemination of data, amplified security and adherence to regulations, and superior decision-making predicated on dependable, prompt, and pertinent information (Dönderler et al., 2005).

1.1.1. Evolution of Data Management Systems

Over time, data management systems have undergone significant advancements to accommodate the increasing quantity, diversity, and speed of data. Data management systems have undergone various stages of evolution that can be broadly classified into different generations (Fry & Sibley, 1976).

1. **First Generation Computers:** The initial stage of computerized data management was characterized by using file-based systems. During the initial stages, data management was predominantly centered on systems based on files. These systems were designed to store information in a structured manner, typically utilizing files or folders arranged hierarchically. They offered rudimentary functionalities for storing and retrieving data. However, they were deficient in established data structures and optimized query capabilities. The challenges encountered in this study were related to the maintenance of data consistency, resulting in information redundancy and inconsistency (Mao et al., 2015).

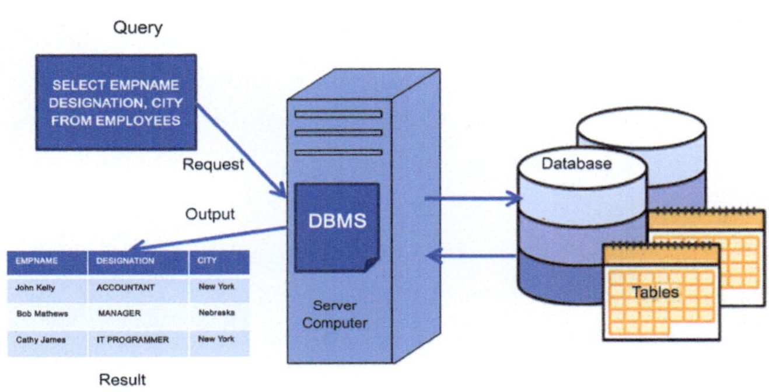

Figure 1.2. Schematic of components of SQL.

Source: Csveda, Creative Commons License.

2. **Second Generation Computers:** The topic of discussion is the second generation of databases, specifically relational databases. The emergence of relational databases represented a noteworthy transition in the realm of data management. Structured data models, which include tables with rows and columns, were introduced by relational databases. These databases also incorporated the application of Structured Query Language

(SQL) for data influence and querying. Relational databases offer superior data organization, credibility, and robust query capabilities. They emerged as the primary systems for managing data for a considerable period (Hurson et al., 1993) (Figure 1.2).

3. **Third Generation Computers:** The emergence of NoSQL databases characterizes the third generation of databases. The advent of big data and novel data formats has brought to light the constraints of relational databases. NoSQL databases have emerged as a viable alternative to traditional SQL databases, providing increased flexibility in data modeling and horizontal scalability. NoSQL databases comprise a range of categories, such as document-oriented, key-value, columnar, and graph records. The system was designed to address the demands of managing semi-structured and unstructured information, facilitating efficient and effective data management at scale.
4. **Fourth Generation Computer:** The fourth generation of databases is characterized by utilizing cloud and distributed databases. The emergence of cloud computing and distributed systems has given rise to a novel cohort of data management systems. Cloud databases utilize the advantageous features of cloud infrastructure, such as scalability, elasticity, and pay-as-you-go simulation, to facilitate the storage and processing of data in a distributed and scalable manner, thus providing organizations with an efficient solution (Oguntimilehin & Ademola, 2014). Distributed databases prioritize the distribution of data, replication, and reliability to facilitate the functioning of distributed applications on a large scale. These technological progressions have facilitated the management of large-scale datasets, instantaneous data analysis, and universal data accessibility.

1.1.2. Challenges and Trends in Data Management

The field of data management has undergone significant advancements; however, it has encountered various obstacles during its progression. Several significant obstacles need to be addressed:

1. **Data Value and Velocity**: The rapid expansion of data has created a demand for organizations to efficiently manage and expedite the processing of substantial amounts of information. Conventional

data management systems may encounter difficulties managing the speed at which data is produced and the magnitude at which it must be stored and evaluated (Lucas-Estañ et al., 2018).

2. **Data Variety and Complexity:** The proliferation of data types, encompassing unstructured, structured, and semi-structured information, has escalated data variety and complexity. The management of heterogeneous data sources, including but not limited to social media, data from sensors, and multimedia files, presents significant difficulties with respect to their integration, storage, and evaluation (Wang et al., 2016).
3. **Data Quality and Integrity:** The maintenance of data precision, uniformity, and integrity continues to pose a significant challenge with regard to data quality. The presence of insufficient or inconsistent data has the potential to result in erroneous insights and decision-making. Implementing data cleansing, validation, and quality assurance procedures is imperative in upholding data integrity.
4. **Data Security and Privacy:** The rising threat landscape and strict regulatory demands have emphasized the necessity for strong measures to ensure data security and privacy. The continuous challenges of safeguarding sensitive data from illicit use, maintaining compliance with privacy rules, and averting data breaches are significant concerns (Yin & Kaynak, 2015).

Trends:

1. **Real-Time Data Processing:** With the increasing demand for real-time insights, data management systems are focusing on handling streaming data and providing real-time analytics capabilities. Technologies such as Apache Kafka and Apache Flink enable processing and analysis of data as it flows in real-time.
2. **DataOps and Agile Data Management:** DataOps practices, inspired by DevOps principles, aim to streamline and automate data management processes. Agile data management methodologies focus on iterative development, collaboration, and continuous delivery of data assets, ensuring faster time-to-value and adaptability (Rodríguez-Mazahua et al., 2016).

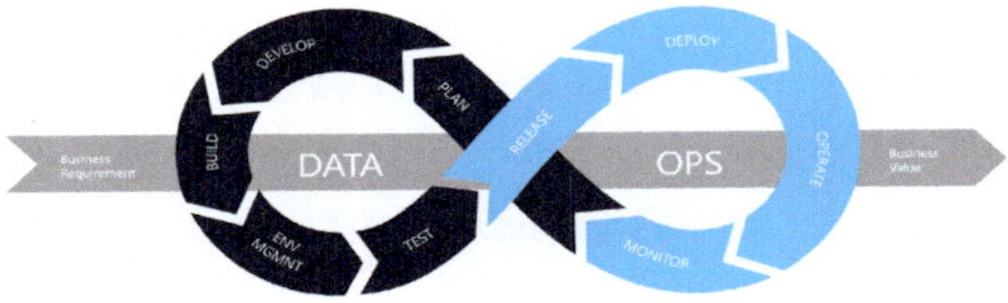

Figure 1.3. *Image of data ops.*

Source: Kent Graziano, Creative Commons License.

3. **Ethical and Responsible Data Management:** Organizations are placing more emphasis on ethical and responsible data management practices. This includes transparent data usage, consent management, and responsible AI practices to ensure fairness, accountability, and transparency in data-driven decision-making.
4. **Data Governance and Compliance:** The establishment of efficient data management systems and the assurance of adherence to regulatory requirements, such as the GDPR and CCPA, can be a challenging task. It is imperative for organizations to clearly delineate data ownership, formulate data governance regulations, and execute procedures for managing the entire data lifecycle (Karim et al., 2020) (Figure 1.4).

Figure 1.4. *Illustration of components of data governance.*

Source: Imperva, Creative Commons License.

Innovations in technology and developing business needs are the driving forces behind current trends in data management. Several noteworthy patterns can be observed, including:

1. **Big Data Analytics:** Organizations are utilizing big data analytics to derive actionable insights from a wide range of extensive and heterogeneous data sources. Sophisticated analytical methodologies, such as machine learning and artificial intelligence (AI), empower enterprises to identify patterns, trends, and predictive models that enhance their decision-making capabilities (Pokorný, 2006).
2. **Cloud-based Data Management:** is a solution that provides benefits such as scalability, flexibility, and affordability for data storage, processing, and analysis. This is made possible through the utilization of cloud computing technology. The utilization of cloud databases and data storage facilities allows organizations to exploit the cloud's capabilities for their data administration requirements, thereby mitigating the responsibility of managing local hardware and software.

In conclusion, data management is crucial for organizations to unlock the full potential of their data assets. The evolution of data management systems has addressed the challenges posed by growing data volumes and diverse data types. Advanced data management encompasses relational databases, NoSQL databases, cloud databases, and distributed databases, each catering to specific use cases (El-Seoud et al., 2017). However, challenges related to data volume, variety, quality, security, and governance persist. Organizations must stay abreast of emerging trends to effectively manage and leverage data for competitive advantage in a rapidly evolving digital landscape.

1.2. ADVANCED DATA MANAGEMENT

1.2.1. Definition and Scope of Advanced Data Management

The term "advanced data management" pertains to using sophisticated methodologies, instruments, and approaches to proficiently manage and exploit substantial amounts of data in intricate and heterogeneous settings (Winkelmann et al., 2017). This approach transcends conventional data management methodologies and centers on tackling the obstacles presented by the continuously expanding data environment. The domain of sophisticated data management encompasses diverse facets such as storing data, data processing, data integration, data analysis, and data governance. The management of data involves a crucial element of data storage, which encompasses the systematic organization, storage, and recovery of data. For a considerable period, conventional relational databases have been the predominant preference for data storage, providing a methodical and tabulated depiction of data. The advent of big information and the necessity to manage heterogeneous data formats has led to the emergence of NoSQL databases as a prevalent substitute (Abadi, 2009). NoSQL databases can accommodate unstructured and semi-structured data due to their adaptability and horizontal scalability.

Advanced data management involves a crucial element of data processing. The process entails the manipulation, alteration, and collection of data with the aim of deriving significant insights. Sophisticated data management systems utilize robust processing capabilities to manage intricate computations and data conversions. The aforementioned tasks encompass data management processes, including data cleansing, normalization, enhancement, and integration. Efficient data processing is a critical factor in producing precise and dependable outcomes (Schmidt et al., 2002).

The integration of data is a crucial component in the realm of sophisticated data management. It is not uncommon for organizations to encounter difficulties in consolidating and analyzing data due to its dispersion across various sources and systems. Using data integration techniques facilitates the synchronization of data originating from diverse sources, culminating in a consolidated perspective that enables exhaustive scrutiny and informed decision-making (Sprague, 1995). ETL processes are widely utilized for extracting data from source systems, standardizing it into a uniform format, and loading it into a designated target system.

The process of analyzing data holds significant importance in the realm of sophisticated data management. The process entails the utilization of statistical, mathematical, and machine-learning methodologies to derive insights and patterns from data. Sophisticated data management systems offer robust analytical capabilities that allow enterprises to conduct descriptive, diagnostic, accurate, and prescriptive analyzes. Using analytics capabilities enables enterprises to make informed decisions based on data, recognize patterns, identify irregularities, and enhance operational efficiencies. Data assets' management, integrity, and security are critical components of advanced data management, which fall under the purview of data governance. Data governance refers to a set of guidelines, protocols, and measures that are implemented to ensure the accuracy, confidentiality, and adherence to the legal requirements of data (Jaradat et al., 2015). The process entails the delineation of data ownership, the establishment of data stewardship positions, the implementation of optimal data management strategies, and the guarantee of data access controls. The implementation of strong data governance protocols is imperative to uphold the accuracy and reliability of data, safeguard confidential information, and adhere to pertinent data privacy statutes (Figure 1.5).

Figure 1.5. Illustration of advance data management.

Source: Datanalysis, Creative Commons License.

1.2.2. Key Concepts and Components of Advanced Data Management

1.2.2.1. Data Modeling and Design

The process of data modeling holds significant importance in the realm of sophisticated data management. The task at hand pertains to the systematic development of the architecture and interconnections of information within a database framework. The primary objective of modeling data is to construct a theoretical and logical depiction of the data, guaranteeing its integrity, uniformity, and usability. Effective storage, retrieval, and tampering operations are made possible by implementing effective data modeling techniques (Urban & Dietrich, 2001).

Within the context of relational databases, the process of data modeling entails the development of entities, characteristics, and relationships, which are frequently depicted through the utilization of entity-relationship diagrams. The process aids in determining the organization of the information, establishing key relationships between primary and foreign entities, and ensuring adherence to regulations governing data integrity. Normalization methods are utilized to eradicate duplicity and guarantee the coherence of data (Smith et al., 2003).

The flexibility of data modeling in NoSQL databases is attributed to the lack of rigid schemas. NoSQL databases facilitate the utilization of dynamic and schema-less structures for data, including documents, pairs of key-value columns, or graphs. Data modeling in databases using NoSQL entails identifying data access structures and the subsequent design of the data structure to align with these patterns. The objective is to enhance the efficiency of data retrieval and adapt to changing data needs.

1.2.2.2. Data Storage and Management

Sophisticated data management systems employ a variety of storage mechanisms to manage substantial amounts of data effectively. Conventional relational databases offer ACID (Atomicity, Consistency, Isolation, Durability) characteristics and facilitate structured data with predetermined schemas. These are extensively utilized in scenarios that necessitate transactional consistency and intricate queries (Schmidt et al., 2002).

Non-relational databases, commonly known as NoSQL databases, provide a more adaptable methodology for storing and handling data. As exemplified by MongoDB, document-oriented databases employ a JSON-like document structure to store data, offering a high degree of adaptability and expandability. Key-value stores, such as Redis, are designed to store data in basic key-value pairs, facilitating rapid and effective data retrieval. As exemplified by Apache Cassandra, column-oriented databases employ a storage architecture that organizes data in columns instead of rows, thereby enhancing both storage and query efficiency (Grolinger et al., 2013). Graph databases such as Neo4j employ a data storage model that utilizes nodes and relationships, thereby enabling intricate graph traversals. The utilization of cloud-based storage solutions has garnered substantial attention in the realm of sophisticated data management. Cloud-based databases, including Amazon Aurora and Google Cloud Spanner, offer data storage solutions that are both scalable and highly available. The aforementioned databases provide features such as automated scaling, replication of data, and integrated disaster recovery mechanisms. Cloud storage offers the benefit of pay-as-you-go pricing, which obviates the necessity for initial infrastructure investments (Schadt et al., 2010) (Figure 1.6).

Figure 1.6. *Image of amazon aurora.*

Source: Carnegie Mellon Database Group, Creative Commons License.

As exemplified by Hadoop Distributed File System (HDFS) and Apache HBase, distributed systems are engineered to manage extensive distributed data storage. Such file systems offer features such as fault tolerance, high reliability, and parallel processing abilities. In large-scale data settings, it is typical to employ them to distribute data across numerous nodes and execute parallel processing (Figure 1.7).

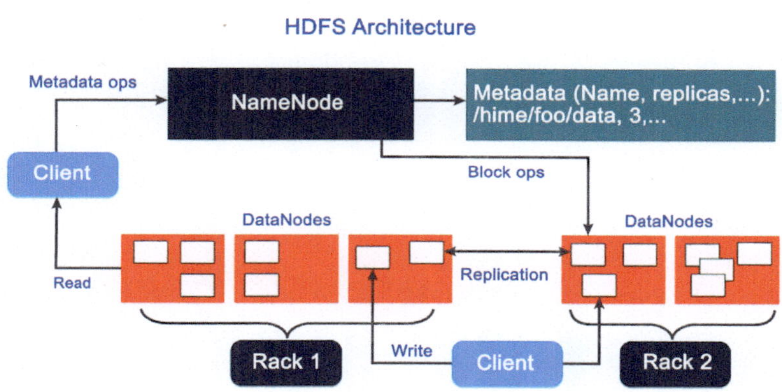

Figure 1.7. Schematic of Hadoop distributed file system (HDFS).

Source: Intellipaat, Creative Commons License.

1.2.2.3. Data Integration and ETL

The integration of data is an essential element in sophisticated data management, facilitating the unification and standardization of data originating from diverse origins. ETL processes are utilized to retrieve data from various sources, standardize it into a uniform format, and subsequently transfer it to a designated database or data warehouse (Sreemathy et al., 2020). The extraction process encompasses collecting information from diverse origins, such as databases, files, APIs, or web scraping. Subsequently, the data undergoes a transformation process aimed at achieving consistency, accuracy, and standardization. These processes encompass data cleansing, standardization, grouping, and enrichment. Ultimately, the modified data is transferred to a designated system, including a data repository, in which it can be subjected to additional scrutiny and employed for various purposes.

There are diverse tools and technologies that can be employed to execute ETL processes. Conventional ETL tools, including Informatica PowerCenter or IBM DataStage, offer visual interfaces for creating and implementing ETL workflows (White, 2005). The platform provides users with various features, including connectors to diverse data sources, transform data functionalities, and scheduling alternatives.

In recent years, the emergence of cloud-based data integration sites such as Apache NiFi or Talend Cloud has facilitated greater flexibility and scalability. The aforementioned platforms provide integration of data as a service, facilitating as a service, and the execution of extract, transform, and load (ETL) procedures by organizations within a cloud-based setting. The service offers pre-configured connectors, data transformation components, and sophisticated data flow management capabilities (Rodzi et al., 2015) (Figure 1.8).

Figure 1.8. *Image of Apache NiFi.*

Source: Wikimedia, Creative Commons License.

1.2.2.4. Data Analysis and Advanced Analytics

Advanced data analysis and sophisticated analytics can be performed using advanced data management methodologies, such as machine learning models, prognostic analytics, and instantaneous analytics. The utilization of these methodologies empowers enterprises to unveil trends, patterns, and conclusions from their information, resulting in enhanced decision-making capabilities, optimized company procedures, and the discovery of novel prospects.

Data mining techniques encompass the systematic exploration and evaluation of extensive datasets with the aim of identifying relationships, trends, and anomalies. The aforementioned techniques encompass clustering, classification, regression, rule mining for associations, and text mining. Data mining techniques, including neural networks, decision trees, and assistance vector machines, are utilized to extract significant patterns and insights from data (Leventhal, 2010). Machine learning algorithms surpass conventional methods for data mining by facilitating systems to acquire knowledge from data and progressively enhance performance. Supervised learning methodologies, such as linear regression and random forests, are employed to perform predictive and classificatory operations. Unsupervised machine learning algorithms, including k-means clustering or analysis of principal components, facilitate the identification of latent structures and trends within the dataset. Reinforcement learning algorithms facilitate the acquisition of optimal actions by systems by providing feedback and rewards (Barton & Court, 2012).

Predictive analytics is a methodology that utilizes statistical techniques and historical data to forecast upcoming occurrences or trends. The process entails constructing predictive models by leveraging past data and utilizing them to anticipate future results or make well-informed judgments. The application of predictive analytics is prevalent across diverse fields such as finance, marketing, healthcare, and production. The field of real-time analytics pertains to examining data as it is received, thereby facilitating prompt decision-making and expeditious action-taking by organizations. The significance of this aspect is particularly noteworthy in sectors where prompt and up-to-date observations hold paramount importance, including finance, e-commerce, or cybersecurity (Berger & Doban, 2014). Analyzing data in motion involves utilizing real-time analytics techniques such as data stream processing, occasion processing, and complicated event processing.

1.2.2.5. Data Governance and Security

The significance of data governance and security is underscored in advanced data management. Data governance refers to a set of guidelines, procedures, and mechanisms that are implemented to guarantee the accuracy, confidentiality, and adherence to regulatory requirements of data. The process entails the establishment of frameworks for data governance, the definition of roles for data stewardship, and the implementation of best practices for data management.

Data governance encompasses the management of data quality, which is responsible for ensuring the precision, comprehensiveness, and uniformity of data (Saed e al., 2008). Data quality procedures, encompassing data profiling, purification, and validation, are implemented to detect and rectify data quality concerns. The management of metadata and data lineage are fundamental constituents of data governance, as they offer insight into the provenance, alterations, and application of data.

Ensuring the security of data is an essential component of sophisticated data management. The process encompasses safeguarding information from unauthorized entry, guaranteeing the confidentiality of data, and adhering to regulations about data protection. Robust data security measures are implemented to protect sensitive data, including restricting access, encryption, tokenization, and anonymization. Data masking techniques can be employed to safeguard privacy by obscuring confidential data while maintaining the usefulness of the data (Trope & Power, 2005). Adherence to data protection protocols, including the General Data Protection Regulation (GDPR) and the California Consumer Privacy Act (CCPA), is a crucial factor in sophisticated data administration. Organizations must establish and enforce policies and procedures that ensure the responsible handling of personal data, afford those with data privacy rights, and mandate the reporting of data breaches when deemed necessary (Begg & Caira, 2012).

1.2.3. Benefits and Advantages of Advanced Data Management Systems

1. **Enhanced Data Accessibility and Usability:** Sophisticated data management systems facilitate efficient access and utilization of data assets, enhancing data usability and accessibility for organizations. By amalgamating and unifying data from diverse origins, users can conveniently retrieve and access pertinent data for analysis and decision-making. The heightened accessibility of data results in enhanced operational efficiency, expedited decision-making processes, and superior business insights (Podgorelec & Kuhar, 2011).
2. **Improved Data Quality and Consistency:** Advanced data management systems implement data quality management methods to guarantee data precision, comprehensiveness, and uniformity, enhancing data quality and consistency. Institutions

can detect and address data quality concerns using data profiling, purification, and validation, leading to more reliable and credible data. Maintaining consistent data across disparate systems and databases is conducive to enhanced integration of information and analysis.

3. **Scalability and Performance:** They are engineered to effectively manage substantial amounts of data and expand their capacity to meet increasing data demands. Organizations can enhance their data management structures to align with their changing needs through horizontal scaling utilizing distributed database systems or vertical scaling through cloud-based solutions (Ren et al., 2014). The ability to scale effectively guarantees the attainment of peak performance and promptness, notwithstanding the growth in data quantities and user requisites.
4. **Agility and Flexibility:** It offers agility and flexibility to accommodate changing company demands and data requirements. Organizations have the flexibility to select the most appropriate data model for their specific use cases, as various database models, including relational and NoSQL databases, are supported. Furthermore, cloud-based solutions can dynamically adjust resources as required, allowing enterprises to respond promptly to variable workloads and evolving data processing requirements.
5. **Improved Decision making and Business Insights:** Organizations can use advanced data analytics and visualization abilities to derive useful knowledge from their data, enhancing decision-making and providing business insights (Shannak et al., 2012). Sophisticated data management systems facilitate investigating and evaluating intricate data collections, revealing latent patterns, tendencies, and associations that might otherwise stay concealed. These insights allow organizations to make informed decisions based on data, recognize potential market prospects, enhance business operations, and attain a competitive advantage (Alavi & Leidner, 1999).

Did You Know?

Advanced data management systems can improve data quality by reducing errors .and inconsistencies by up to 50%, leading to more reliable and accurate insights

The field of advanced data management comprises several fundamental principles and constituents that facilitate the efficient management and utilization of substantial amounts of data across various settings for organizations. Advanced data management systems offer a multitude of advantages by integrating data modeling, retention, integration, evaluation, governance, and security. These benefits include heightened access to data, enhanced data quality, scalability, flexibility, and the capacity to extract valuable business insights (Chidwick, 1996). In the contemporary data-driven landscape, organizations must adopt sophisticated data management strategies to fully leverage their data resources and achieve a competitive edge, given data's escalating intricacy and magnitude.

1.3. RELATIONAL DATABASES

1.3.1. Overview of Relational Database Management Systems (RDBMS)

Relational Database Management Systems (RDBMS) are extensively utilized data management systems in various industries. A structured methodology is employed to arrange and maintain data in tables that possess predetermined associations. RDBMS provide robust capabilities and functionalities that facilitate efficient storage (Malecha et al., 2010). The relational model is a fundamental concept in the realm of RDBMS. It involves the organization of data into tables that are comprised of rows and columns. In the context of data modeling, it can be observed that every table represents an entity or an association between entities (Grad, 2012). Furthermore, the columns within these tables define the various attributes or properties associated with said entities. The establishment of relationships between tables is facilitated by utilizing keys, including the main and foreign keys.

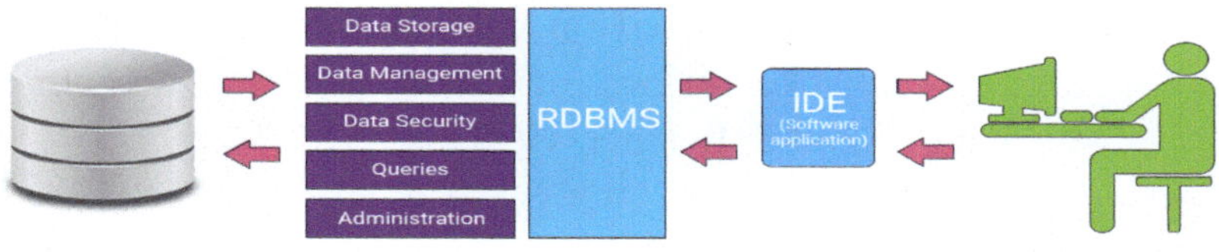

Figure 1.9. Schematic of relational database management systems (RDBMS).

Source: Vetrabelo, Creative Commons License.

RDBMS offer robust mechanisms for data integrity and guarantee the stored data's accuracy and consistency. Data constraints, including uniqueness, referential security, and data types, are enforced to preserve the confidentiality of the database. The ACID properties are essential in ensuring data's dependability and transactional consistency (Zhang et al., 2001) (Figure 1.9).

1.3.2. Data Modeling and Design in Relational Databases

The process of data modeling plays a pivotal role in the development of a relational database. The process entails the identification of entities, relationships, and characteristics that accurately depict the real-world domain being depicted. ER diagrams are a widely utilized tool for graphically illustrating the configuration of a database, depicting the entities, their respective attributes, and the interconnections among them (Bordoloi & Kalita, 2013). Normalization is a crucial element of data modeling within the context of relational databases. Data normalization is a systematic approach to organizing data to reduce redundancy and guarantee data integrity. Normalization techniques, including the normalization forms such as the first normal form and the second normal form, aid in eradicating data anomalies and guaranteeing effective retention and retrieval.

In designing a database, it is crucial to identify the suitable main and foreign keys. Primary keys serve as distinct identifiers for individual rows within a given table, whereas foreign keys facilitate the establishment of connections between tables by referring to primary keys from other tables. It is imperative to exercise caution when defining and preserving these keys to uphold the integrity of the data.

1.3.3. SQL and Advanced Querying Techniques

SQL is widely recognized as the standard language for communicating with relational databases. The software offers a diverse range of commands and activities that facilitate the execution of multiple tasks, such as retrieving information, insertion, modification, and deletion. SQL statements can be classified into two main categories: Data Definition Language (DDL) and Data Manipulation Language (DML).

DDL statements are utilized to establish the arrangement and structure of the database (Giakoumakis & Galindo-Legaria, 2008). The software permits the creation of tables, the establishment of constraints, and the administration of indexes. In contrast, DML statements are employed to alter and retrieve data. The aforementioned commands encompass SELECT, INSERT, UPDATE, and DELETE.

Apart from fundamental SQL operations, the utilization of sophisticated querying techniques can substantially amplify the adaptability and potency of data retrieval. The aggregate functions, including SUM, COUNT, AVG, and MAX/MIN, facilitate the computation of values on data sets. Join operations are utilized to merge data from various tables by establishing predetermined relationships. The utilization of subqueries enables the incorporation of one query within a different one, thereby facilitating the retrieval of data with increased complexity and sophistication (Imam et al., 2018). The RDBMS provides significant value by providing views and stored procedures. Views are a type of virtual tables that are created from the original data source. They offer a simplified and regulated perspective of the data to the end-users. Stored procedures are a predetermined collection of queries in SQL that can be carried out as a cohesive unit. The implementation of these features helps augment the modularity, reuse, and security of database operations.

1.3.4. Performance Tuning and Optimization in RDBMS

Efficient and adaptable data management can be achieved by optimizing the efficiency of a RDBMS. There exist various methodologies that can be utilized to optimize the efficiency of a database system (Woods et al., 2014).

The utilization of indexing is a widely employed methodology to enhance the efficiency of query processing. Indexes are a type of data structure that facilitates expedited data retrieval by generating a duplicate of a particular subset of data and arranging it to enable swift searching based on designated columns.

The appropriate selection and construction of indexes on frequently queried columns can enhance the efficiency of data retrieval operations. Optimizing queries is a crucial component of enhancing system performance. RDBMS employ query optimizers to scrutinize the execution plans of queries and ascertain the most optimal approach to retrieve data.

Various methods, including cost-based optimization, query rewriting, and query hints, can be utilized to enhance and optimize the performance of queries.

The optimization of resource utilization and enhancement of overall system performance can be achieved by database administrators through the tuning of system parameters, including but not limited to buffer sizes, cache arrangements, and processing speed settings. The consistent monitoring, profiling, and analysis of system performance metrics aid in the identification of bottlenecks and optimization needs (Sharma et al., 2016).

Moreover, the utilization of denormalization can be strategically applied in specific situations to enhance system efficiency. Denormalization is a deliberate strategy in database design that entails the introduction of redundancy by copying data or calming down normalization rules.

The utilization of this methodology has the potential to reduce the cost of join operations and enhance the efficiency of queries in settings that prioritize reading operations.

To sum up, relational databases are extensively utilized due to their strength, data consistency, and querying proficiency. Effective utilization of relational database management systems necessitates adeptness in data modeling and design, advanced querying techniques, and optimization strategies for performance tuning. By utilizing these methodologies, entities can attain proficient and successful data management resolutions that fulfill their requirements (Bhushan et al., 2016). The present discourse aims to provide an overview of NoSQL databases.

1.4. NOSQL DATABASES

The popularity of NoSQL (Not Only SQL) databases has increased considerably in recent times owing to their capacity to manage substantial amounts of unstructured or semi-structured data while offering scalability, flexibility, and superior performance. In contrast to conventional relational databases, NoSQL databases present an alternative data storage methodology and data model. These systems are engineered to fulfill the demands of contemporary applications that necessitate the storage and manipulation of extensive quantities of data (Kemper et al., 1999).

Non-relational databases, commonly referred to as NoSQL databases, are frequently employed in situations where conventional relational databases encounter difficulties, such as managing large volumes of data, performing real-time analytics, and operating distributed systems. These are especially well-suited for use cases that necessitate horizontal scalability, wherein data is dispersed across numerous nodes or clusters.

1.4.1. Types and Categories of NoSQL Databases

Various classifications and varieties of NoSQL databases exist, each possessing unique attributes and advantages (Strauch et al., 2011). The principal classifications of NoSQL databases comprise document-oriented databases, key-value stores, column-family databases, and graph databases.

1. **Document Databases:** These are a type of database that follows a document-oriented format for storing and retrieving data. This format usually employs JSON (JavaScript Object Notation) or XML. It is possible for documents to exhibit a hierarchical structure comprising nested key-value pairs. The malleable nature of this data model facilitates the facile management of intricate and diverse data configurations. Document-oriented databases offer robust querying functionalities like indexing, full-text search, and grouping. MongoDB, Couchbase, and Elasticsearch are among the commonly used document databases.
2. **Key-Value Databases:** They organize data in the form of key-value pairs. The key serves as a distinctive identifier for individual values, and the database facilitates effective retrieval and alteration operations predicated on the key. Key-value databases are a viable option for caching, managing sessions, and basic data storage use cases due to their ability to provide high levels of performance and scalability. Redis, Riak, and Amazon DynamoDB are among the key-value databases that are commonly used (Tudorica & Bucur, 2011).
3. **Columnar databases:** They arrange data column-wise rather than the traditional row-wise organization. Using columnar storage methodology facilitates effective compression and regaining of particular columns, rendering it highly appropriate for analytical tasks and information warehousing. Columnar databases are proficient in managing extensive amounts of data and executing collection and analytics queries. Apache Cassandra, Apache HBase, and Vertica are widely recognized as prominent examples of columnar databases (Chen et al., 2019).
4. **Graph Databases:** They utilize nodes, edges, and characteristics to store and represent data. This structure enables the efficient storage and navigation of connections among entities. Graph databases are proficient in managing extensively interconnected data, such as those found in social networks, and systems for recommending and identifying fraudulent activity. Graph-based querying languages are offered as a powerful tool for the purpose

of navigating and analyzing intricate relationships. Prominent graph database examples include Neo4j, Amazon Neptune, and Apache Giraph.

1.4.2. Data Modeling and Design Considerations for NoSQL Databases

The process of data modeling in NoSQL databases necessitates a distinct methodology in contrast to that of relational databases (Tauro et al., 2012). The dynamic schema designs of NoSQL databases enable flexibility, wherein each record can possess distinct fields and structures, owing to the absence of a rigid schema. The following are essential factors in modeling within NoSQL databases:

1. **Denormalization:** It is a common practice in NoSQL databases, whereby intentional duplication of data is introduced to enhance data retrieval efficiency and minimize the need for joins. The act of denormalization provides support for the optimization of query execution and serves to diminish the necessity for intricate join operations.
2. **Data Access Pattern:** Understanding the data access patterns of an application is imperative to devise an effective data model. To achieve optimal performance, it is recommended that NoSQL databases are tailored to match the particular read and write behaviors of the application (Atzeni et al., 2020).
3. **Scalability and Distribution:** NoSQL databases are engineered to achieve horizontal scalability by distributing data across numerous nodes or clusters. The process of data modeling requires careful consideration of data division and sharding methods to ensure equitable distribution of data and efficient execution of queries.
4. **Consistency Models:** NoSQL databases provide various consistency models, including eventual consistency, eventual strong consistency, and ultimately strong consistency. The selection of an appropriate consistency model is contingent upon the particular demands of the application and the balancing performance and data uniformity considerations (Bugiotti et al., 2014).

1.4.3. Querying and Data Manipulation in NoSQL Databases

Non-relational (NoSQL) databases offer distinct methods for querying and manipulating data compared to conventional SQL-based databases. Although individual categories of NoSQL databases may employ distinct query languages or APIs, certain shared patterns and techniques exist for querying and influencing data in NoSQL databases.

1. **CRUD Operations:** NoSQL databases facilitate the execution of CRUD operations, which involve creating, retrieving, modifying, and deleting data. The execution of these operations is enabled by the APIs and query languages that are offered by each NoSQL database (Yubin et al., 2013).
2. **Querying with Non-SQL Languages:** The retrieval and manipulation of data in NoSQL databases may involve the use of particular query languages or APIs, depending on the database category. Document-oriented databases frequently employ a versatile query language, such as MongoDB's query language, which enables data filtration, ordering, and selection.
3. **MapReduce and Aggregation:** Numerous NoSQL databases facilitate using MapReduce and aggregation structures for executing intricate data conversions and analytics. The utilization of these frameworks facilitates the decentralized handling of voluminous data sets over numerous nodes or clusters.
4. **Secondary Indexing:** NoSQL databases offer secondary indexing techniques to facilitate optimized retrieval of particular fields or attributes in the data (Gunawan et al., 2019). The implementation of secondary indexes can have a notable impact on the efficiency of queries as it enables expedited a link to the indexed data.

To conclude, NoSQL databases have surfaced as potent substitutes for conventional relational databases, providing scalability, flexibility, and performance benefits.

Comprehending the various categories of NoSQL databases, their pertinent data modeling factors, and querying functionalities is imperative for proficiently utilizing these databases in sophisticated data management systems. By aligning data models and design choices with application-specific demands, developers and data professionals can effectively leverage the complete potential of NoSQL databases in managing contemporary data-intensive workloads.

Tip

Domestic robot is a type of service robot, an autonomous robot that is primarily used for household chores, but may also be used for education, entertainment or .therapy

1.4.4. Cloud Computing and Its Impact on Data Management

The advent of cloud computing has significantly transformed how entities store, handle, and manipulate their data (Agrawal et al., 2011). Cloud computing is a service delivery model that entails providing computing resources, including databases, storage, and applications, through the Internet on a pay-per-use basis. The migration from conventional on-premises infrastructure to cloud computing has significantly influenced data management methodologies.

Cloud computing presents various benefits in the realm of data management. Initially, it offers nearly boundless scalability and elasticity, enabling entities to rapidly adjust the magnitude of their storage and computational resources in accordance with their requirements. This obviates the necessity for initial infrastructure expenditures and facilitates nimble data administration.

Moreover, cloud computing facilitates commonplace access to data from any location, thereby promoting collaboration and work from home (Rallapalli et al.,2016).

The software offers smooth and uninterrupted integration with various cloud-based services, including data analysis and machine learning systems, thereby facilitating sophisticated data analysis and processing. Cloud databases are essential in leveraging the advantages of cloud computing for effective data administration.

1.4.5. Cloud Database Architectures and Deployment Models

Cloud databases refer to specialized database management systems (DBMS) specifically designed to operate within the cloud computing environment. The cloud-based solutions provided by the company are characterized by their adaptability and capacity for expansion, enabling efficient data storage and management. The design and deployment models of cloud databases exhibit variability contingent upon specific needs and needs of organizations. The following are frequently employed models:

1. **Database-as-a-Service (DBaaS) Model:** It involves the complete management of the database infrastructure, encompassing hardware, software, and maintenance, by the cloud service provider. Organizations may prioritize the utilization and administration of data without being concerned with the foundational infrastructure (Malliga, 2012).
2. **Virtual Machine-Based Deployment:** It is an approach favored by certain organizations, whereby they opt to deploy their own virtual machines in the cloud and install conventional DBMS software on these virtual machines. This confers greater autonomy to the users in managing the database environment, albeit at the cost of increased managerial exertion.
3. **Serverless Databases:** It represent a contemporary paradigm in which the cloud provider is responsible for managing the database facilities and dynamically adjusting resource allocation in response to fluctuations in demand (Holden et al., 2011). The billing system is based on actual usage, rendering it a financially efficient alternative for specific workloads.
4. **Distributed Database Architecture:** Cloud databases can facilitate distributed database architectures that extend across various geographical regions or data centers. The aforementioned feature facilitates increased availability, mitigation of catastrophic events, and streamlined duplication of data to enhance operational efficiency and dependability.

1.4.6. Benefits and Challenges of Using Cloud Databases

Cloud-based databases provide a multitude of advantages that facilitate improved data management methodologies:

1. **Scalability:** The scalability of cloud databases allows for effortless expansion or contraction in response to changes in data volume and patterns of use. The property of elasticity guarantees that enterprises can effectively manage increasing data requirements without encountering any interruptions or limitations in capacity.
2. **Cost-Efficiency:** Pay-as-you-go cloud databases do away with the need for significant up-front hardware and software investments.

Figure 1.10. Image of cloud databases.

Source: Database, Creative Commons License.

3. **Accessibility and Collaboration:** Cloud databases facilitate convenient and protected data accessibility from any location, thereby promoting collaboration among geographically dispersed teams. The facilitation of accessibility fosters prompt decision-making and expedites the process of innovation (Al Shehri, 2013).
4. **High Availability and Fault Tolerance:** Cloud-based databases frequently offer integrated functionalities for ensuring high availability, data replication, and recovery from disasters, thereby enhancing fault tolerance. The practice of replicating data across several data centers serves to mitigate the potential for data loss or system downtime (Figure 1.10).

1.4.7. Challenges

1. **Data Integration:** The process of integrating cloud databases with on-premises systems can be intricate, necessitating meticulous planning and execution to guarantee uninterrupted data flow and synchronization (Zhang et al., 2019).
2. **Performance and Latency:** The utilization of cloud databases can potentially result in the introduction of network latency, particularly in instances where data retrieval is being conducted from distant locations. To achieve optimal performance, it is necessary to take into account architectural factors and effectively employ caching and indexing methodologies.

3. **Vendor Lock-in:** The issue of vendor lock-in is a significant concern for organizations that utilize cloud databases, as they must exercise prudence in this regard. The evaluation of portability and compatibility alternatives is imperative in order to circumvent reliance on a particular cloud vendor (Anandhi & Chitra, 2012).

1.4.8. Data Security and Privacy in the Cloud

Data confidentiality and integrity preservation represent a paramount issue in utilizing cloud-based databases (Samarati, 2014). Although cloud providers allocate significant resources toward security measures, it is incumbent upon organizations to assume accountability for safeguarding their data. The following factors should be taken into account:

1. **Encryption:** Encrypting data while it is at rest and in movement is of utmost importance in safeguarding confidential information. Cloud-based databases frequently offer encryption alternatives, which organizations ought to utilize to guarantee the confidentiality of their data.
2. **Access Controls:** The implementation of strong access controls, such as multi-factor authentication (MFA) and role-based access, can effectively mitigate the risk of unauthorized access to sensitive data and cloud databases (di Vimercati et al., 2019).
3. **Compliance:** In order to effectively store and manage data in the cloud, organizations must possess a comprehensive understanding of pertinent privacy laws and industry standards and to ensure compliance with them. This could potentially entail adherence to regulations regarding the physical location of data storage, protocols for notifying individuals in the event of a data breach, and the establishment of policies governing the handling of sensitive information.
4. **Data Backups and Disaster Recovery:** The implementation of data backups and recovery strategies is crucial in reducing the likelihood of data loss and maintaining uninterrupted business operations in the case of a cloud outage or other types of interruption (Esposito et al., 2018).

It is recommended that organizations implement an extensive safety plan maintain a strong security posture in the cloud. This strategy should include typical security audits, vulnerability evaluations, and employee training.

To sum up, cloud-based databases have revolutionized the way data is managed by offering solutions that are scalable, cost-efficient, and easily accessible. The aforementioned benefits encompass flexibility, accessibility, and high accessibility.

Notwithstanding, entities are required to tackle obstacles such as data integration and performance optimization (Kong et al., 2018). Furthermore, it is imperative to meticulously execute data security and privacy protocols to protect confidential data stored in the cloud. Through a comprehensive comprehension of the complexities inherent in cloud-based databases and the adept handling of these factors, entities can harness the potential of cloud technology to augment their data management proficiencies.

1.5. DISTRIBUTED DATABASES

A distributed database refers to numerous linked databases geographically dispersed across various physical locations. Distributed databases present several benefits in contrast to conventional centralized databases, including heightened fault tolerance, increased scalability, and improved performance (Rothnie Jr et al., 1980). This section will explore the basic ideas and concepts underlying distributed databases.

Comprehending the notion of data distribution is imperative in gaining a comprehensive understanding of distributed databases. In a distributed database, information is partitioned and distributed among numerous nodes or servers, which may be located in different geographical locations. The distributed nature of this arrangement enables streamlined data retrieval and access, as queries can be concurrently executed throughout the distributed nodes (Chung, 1990).

1.5.1. Concepts of Data Distribution, Replication, and Consistency

1. **Data Distribution:** This process entails partitioning a database into distinct subsets and allocating each subset to a designated node or server within the distributed system. Various methodologies exist for data distribution, such as horizontal partitioning, which involves dividing tables based on rows, and vertical partitioning, which involves dividing tables based on columns (Breitbart & Korth, 1999). The determination of the appropriate strategy for data distribution is contingent upon various factors, such as the

manner in which queries are executed, the patterns of data access, and the specifications of the system.

2. **Data Replication:** It refers to the procedure of generating and preserving numerous duplicates of information throughout various nodes within a dispersed database. The process of replication has been observed to have a positive impact on the resilience of a system against faults, as well as on the accessibility of data. Diverse replication techniques exist, including complete replication, which entails the replication of all data to every node, and partial replication, which involves the replication of only subsets of data (Milani & Navimipour, 2016). Formulating replication strategies necessitates achieving equilibrium among the factors of consistency, efficiency, and data synchronization.
3. **Consistency:** The concept of consistency in a database that is distributed pertains to the maintenance of a coherent state across all copies of data. The attainment of uniformity in distributed systems poses a formidable challenge due to network latency, concurrency, and failures of nodes (Goel & Buyya, 2007). Consistency models, which include rigid, eventual, and related consistency, establish the degree of synchronization and cooperation among distributed nodes. Consistency models play a crucial role in preserving the integrity of data and guaranteeing consistent outcomes of queries.

1.5.2. Distributed Query Processing and Optimization

Efficient processing and optimization of queries are essential in distributed databases to facilitate effective data retrieval while minimizing network overhead. The process of distributed query processing entails breaking down user queries into smaller sub queries and allocating them to suitable nodes for execution. The last search result is obtained by aggregating the outcomes from individual nodes.

1. **Query Decomposition:** It refers to the process of dividing user queries into fewer sub queries that can be carried out separately on various nodes (Yu & Chang, 1984). The decomposition process considers various factors, such as the distribution of data, replication, and dependencies of queries, to ascertain the optimal execution location for each sub query.

2. **Query Optimization:** It is a process that strives to reduce the total response time of a query by identifying and selecting the most effective plan for executing the query. The optimization of queries in a distributed database is a more intricate process, as it is influenced by various factors such as data disintegration, data distribution, and the costs associated with network communication. Optimizing involves utilizing cost-based evaluation, heuristic algorithms, and statistical methods to approximate the most optimal plan of execution.

1.5.3. Scalability and Fault Tolerance in Distributed Databases

The ability to handle substantial amounts of data and endure system failures are fundamental attributes of distributed databases, commonly called scalability and fault tolerance (Bagui & Nguyen, 2015).

1. **Scalability:** The horizontal scalability feature of distributed databases allows adding extra nodes to the system to cater to the growing data storage and processing demands. The attainment of scalability can be facilitated by employing methodologies such as data division, load balancing, and dispersed file systems. Distributed databases can efficiently handle large volumes of data and high transaction prices by distributing the workload and data across multiple nodes (Abdelhafiz & Elhadef, 2021).
2. **Fault Tolerance:** It is a crucial aspect of distributed databases as it guarantees the system's continuous operation despite failures of nodes or network disruptions. The implementation of fault tolerance can be facilitated by utilizing various techniques such as data replication, redundant operation, and dispersed consensus protocols, including Paxos and Raft. In the event of a node failure, the system can sustain uninterrupted operation by utilizing replicated data copies stored on alternative nodes (Kanwar et al., 2013).

1.6. DATA INTEGRATION AND ETL

1.6.1. Data Integration and Its Importance

The amalgamation of data from multiple sources into a cohesive and uniform perspective is referred to as data integration. In the contemporary era of data-centricity, entities are confronted with copious amounts of data originating from diverse systems, databases, and kinds. The integration of data is a pivotal factor in facilitating organizations to acquire significant insights, arrive at informed decisions, and extract significant knowledge from their data resources.

The necessity for a comprehensive data perspective is a fundamental rationale behind the significance of data integration (Golshan et al., 2017). It is common for organizations to possess data that is dispersed across various systems, encompassing relational databases, NoSQL databases, legacy systems, and external sources, such as APIs and web services. By integrating diverse data sources, entities can attain a consolidated perspective of their data, eradicating isolated data repositories and facilitating extensive examination and documentation. The process of integrating data also serves to enhance the consistency and precision of the data. The distribution of data across multiple systems increases the likelihood of encountering inconsistencies, duplications, and discrepancies within the data. Through data integration, organizations can effectively ensure the integrity and consistency of their data, thereby mitigating the likelihood of errors and furnishing dependable information to make informed decisions (Linthicum, 2017).

Moreover, integrating data facilitates the extraction of the inherent worth of data for enterprises through the provision of sophisticated analytics and business intelligence (BI). By integrating data from diverse sources, entities can discern patterns, tendencies, and correlations that might otherwise stay concealed. Consequently, this facilitates them to arrive at better-informed judgments, streamline procedures, and attain a competitive advantage in the marketplace (Zhou, 2016).

1.6.2. Extract, Transform, Load (ETL) Processes and Techniques

The process of ETL is a commonly utilized method for integrating data. The process comprises three primary stages: extraction, transformation, and loading (Vassiliadis, 2009).

1. **Extraction:** The initial stage of data collection, known as extraction, entails the acquisition of information from various sources. Possible academic rewrite: This may involve accessing data from various sources such as databases, files, application programming interfaces (APIs), web scraping, or other pertinent repositories. The methodology for extracting data may exhibit

variability contingent upon its underlying systems and formats. Data extraction can be accomplished through various methods, such as direct querying, file transfers, and specialized tools.

2. **Transformation:** It is a crucial step in the data processing pipeline, as it enables the attainment of the extracted data's consistency, quality, and compatibility. The data transformation process encompasses diverse procedures, including but not limited to data cleansing, filtering, aggregation, enhancement, and data type conversions (Seenivasan, 2023). The objective is to establish a uniform format for the data, address any discrepancies, and make it ready for assimilation into the intended system.

The methods of transformation may differ based on the particular specifications and characteristics of the data. This may encompass various tasks such as conducting computations, implementing commercial regulations, consolidating data sets, or even assimilating external data resources. Data integration places and ETL (Extract, Transform, Load) tools offer a variety of features to facilitate these transformation processes.

3. **Loading:** The loading phase pertains to the process of transferring the modified data into the designated database or target system. The potential destination systems may encompass a data warehouse, a reporting database, or any other analogous platform. The process of loading data guarantees that it is arranged, indexed, and optimized to facilitate effective querying and analysis (Bansal & Kagemann, 2015). The loading process may vary depending on the intended system, encompassing bulk loading, incremental loading, or real-time data streaming.

1.6.3. Tools and Technologies for Data Integration and ETL

A plethora of tools and technologies are available to facilitate and optimize data integration and ETL processes. These tools offer capabilities for extracting, transforming, and loading data, as well as provisions for managing data quality, planning, monitoring, and error handling. The discipline of data integration and ETL commonly employs a variety of tools and technologies.

1. ETL tools are specialized software applications intended to aid in the seamless integration of data and the execution of ETL (Extract, Transform, Load) procedures (Sreemathy et al., 2021). The aforementioned tools offer graphical user interfaces and intuitive drag-and-drop features for the purpose of creating and

overseeing ETL workflows. Prominent ETL tools comprise Informatica PowerCenter, IBM InfoSphere DataStage, Microsoft SQL Server Integration Services (SSIS), and Talend Open Studio.

2. **Data Integration Platforms:** It provide extensive solutions for integrating data, including ETL and governance of data, data quality, and data enhancement. The aforementioned platforms offer a consolidated setting for overseeing data integration workflows, supervising data streams, and guaranteeing data coherence. Prominent data integration platforms comprise Talend Data Fabric, Oracle Data Integrator, and SAP Data Services.
3. **Change Data Capture (CDC) Tools:** They are utilized to capture and monitor alterations that have been made to the source data (Sreemathy et al., 2021; Chakraborty et al., 2017). This enables organizations to selectively integrate and replicate the updated information solely into the target system. The aforementioned tools facilitate data integration in real-time or near real-time, thereby ensuring synchronization between the source and target systems. The tools utilized by the CDC comprise Oracle GoldenGate, Attunity Replicate, and HVR Software (Figure 1.11).

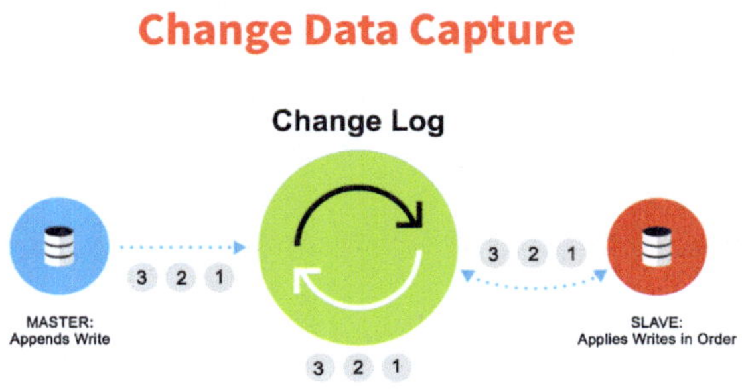

Figure 1.11. Image of change data capture (CDC).

Source: Sarad Mohanan, Creative Commons License.

1.6.4. Challenges and Best Practices in Data Integration

The process of integrating data can pose several obstacles that organizations must overcome to achieve favorable outcomes in their integration endeavors. Several prevalent difficulties include:

1. **Data Quality and Consistency:** The issue of data consistency and accuracy arises due to the possibility of variations in the quality levels, discrepancies, or mistakes in data obtained from diverse sources. The integration process necessitates the utilization of data purification, standardization, and validation methods to guarantee data quality and consistency.
2. **Data Security and Privacy:** The preservation of data security and privacy is crucial when amalgamating data from various sources. Organizations must implement suitable measures to safeguard confidential information, comply with regulatory mandates, and ensure adequate access controls (Gligorijević & Pržulj, 2015).
3. **Scalability and Performance:** The criticality of scalability and performance increases as data volume and complexities expand. Efficient handling of large datasets and optimization of performance to reduce processing times should be the primary considerations in the design of data integration processes.

In order to surmount these challenges and guarantee efficacious data integration, institutions ought to adhere to optimal methodologies, which encompass the following:

1. **Data Implementation:** The implementation of a framework for data governance is instrumental in establishing data standards, regulations, and processes for data integration (Nargesian et al., 2022). The integration process is facilitated by ensuring the accuracy of data, consistency, and compliance.
2. **Meta-data Management:** Effective management of metadata pertaining to the amalgamated data is of utmost importance in comprehending data lineage, terms, and business rules. Effective metadata management facilitates streamlined data exploration, documentation, and long-term upkeep.
3. **Incremental and Agile Approach:** Adopting a gradual and agile strategy for data integration enables organizations to provide early value and adjust to evolving demands. Dividing the integration procedure into feasible iterations mitigates potential hazards and facilitates expedited comprehension (Nargesian et al., 2023; Budgen et al., 2007).
4. **Testing and Validation:** The rigorous examination and verification of integrated data are imperative to detect and correct any inconsistencies, inaccuracies or gaps. Data validation

techniques are imperative in guaranteeing consolidated data's precision and dependability.

Through the implementation of effective strategies and adherence to established methodologies, entities can attain favorable outcomes in the realm of data integration. This achievement facilitates the ability to utilize data as a basis for informed decision-making, enhance the efficacy of operational processes, and attain a competitive edge in the contemporary digital milieu.

The integration of data is a crucial component in facilitating the ability of organizations to extract the full potential of their data resources (Ehwerhemuepha et al., 2020). The ETL (Extract, Transform, Load) process is facilitated by a range of tools and technologies, enabling the smooth transfer, conversion, and integration of data into designated systems. Although data integration presents certain difficulties, adhering to optimal methodologies and taking into account aspects such as data integrity, confidentiality, and expandability can result in efficacious integration endeavors. By utilizing sophisticated data management methodologies, entities can harness the potential of consolidated data to attain significant discernments, stimulate originality, and flourish in the epoch of SQL, NoSQL, cloud, and dispersed databases.

1.7. ADVANCED DATA ANALYTICS

In the contemporary era of data-centricity, enterprises persistently pursue valuable insights from their extensive data reservoirs to facilitate informed decision-making and attain a competitive advantage. The present chapter is a preliminary exposition to sophisticated data analytics, delving into the fundamental principles, techniques, and methods that empower institutions to derive valuable insights from their data (Huttunen et al., 2019; Mitchell, 1999).

The term "data analytics" pertains to the systematic examination of extensive data sets with the aim of discovering recurring patterns, tendencies, and valuable insights that can inform and guide business strategies and decisions. Advanced data analytics surpasses fundamental descriptive analysis and explores more intricate methodologies for extracting significant insights. The process entails utilizing diverse techniques and instruments to acquire a more profound understanding, formulate forecasts, and facilitate decision-making based on empirical evidence.

1.7.1. Data Mining and Machine Learning Techniques

Data mining is considered a fundamental cornerstone of advanced data analytics. The process of data mining entails the identification of patterns, relationships, and anomalies within extensive datasets through the utilization of statistical and machine learning algorithms (Bose & Mahapatra, 2001). By implementing diverse data mining methodologies, enterprises can reveal concealed patterns and acquire significant insights that can be utilized for multiple objectives, including customer segmentation, fraud identification, and market evaluation.

The utilization of machine learning methodologies is of utmost importance in the realm of sophisticated data analytics. Machine learning algorithms enable computational systems to learn from data and develop predictions and explicit programming. Classification and regression tasks are commonly addressed through the utilization of supervised learning algorithms, including decision trees and neural networks. Unsupervised learning methodologies, like association and clustering rule mining, facilitate the detection of latent patterns and interdependencies within the dataset (Teng & Gong, 2018; Hayn et al., 2018). In contrast, reinforcement learning algorithms facilitate the acquisition of knowledge by machines through the process of experimentation and feedback.

1.7.2. Predictive Analytics and Data-Driven Decision Making

Predictive analytics is a subfield of sophisticated data analytics that centers on forecasting forthcoming results by utilizing past data and statistical models. Through the examination of past patterns and associations, institutions can construct anticipatory models that prognosticate forthcoming occurrences or conducts. The aforementioned models possess the capability to forecast customer behavior, enhance inventory management, anticipate equipment malfunctions, and alleviate potential hazards.

The process of data-driven decision making entails making informed decisions by analyzing and deriving insights from data. Organizations can transition from decision making based on intuition to a reliance on evidence and facts by utilizing sophisticated data analytics techniques (Milosevic et al., 2016). The utilization of data-driven decision making empowers organizations to enhance their processes, recognize potential opportunities, reduce risks, and attain a competitive edge.

1.7.3. Real-time Analytics and Streaming Data Processing

The analysis of data as it is generated, commonly known as real-time analytics, is an essential aspect of advanced data analytics. It enables organizations to make informed decisions and take proactive measures promptly. The application of real-time analytics is especially pertinent in sectors that include finance, healthcare, and the Internet of Things (IoT), as prompt insights can result in notable benefits.

Real-time analytics relies on the fundamental concept of streaming data processing. The process entails the real-time processing and analysis of data streams, the extraction of insights, and initiating actions according to the incoming data (Koltay, 2016). The utilization of complex event processing (CEP) and stream processing structures has enabled organizations to effectively manage and analyze high-velocity data streams originating from diverse sources (Figure 1.12).

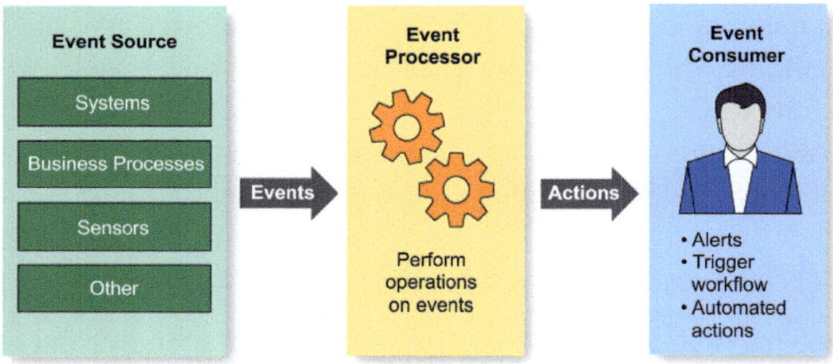

Figure 1.12. Diagram of complex event processing (CEP).

Source: Hazelcast, Creative Commons License.

Through real-time analytics and streaming information processing, entities can promptly identify irregularities, oversee crucial systems, enhance operations, and furnish customized experiences to their clientele.

1.8. DATA GOVERNANCE AND DATA QUALITY

In contemporary times, the practice of utilizing data to inform decision-making has become prevalent. In this context, data governance serves as a crucial mechanism for safeguarding data's dependability, authenticity, and confidentiality within sophisticated data management frameworks.

The concept of data governance pertains to formulating and implementing regulations, procedures, and principles that facilitate the efficient management of data resources (Khatri & Brown, 2010; Weber et al., 2009). In light of the escalating quantity and intricacy of data, organizations must prioritize data governance to attain optimal outcomes in data management.

The importance of data governance is paramount in the realm of sophisticated data management due to various factors. Initially, it furnishes a structure for managing data throughout its entire lifespan, commencing from the point of acquisition to the point of disposal. Organizations can ensure responsible data management in compliance with legal and regulatory requirements by implementing measures such as creating obvious ownership, accountability, and data stewardship. Implementing efficient data governance practices can facilitate transparency and instill confidence in data, enabling organizations to make well-informed decisions grounded in dependable and precise information.

Moreover, the implementation of data governance facilitates the ability of organizations to sustain uniformity and conformity of data throughout diverse systems and departments. Organizations may develop a shared comprehension of data elements and their significance by describing data standards, standard names, and classifications (Al-Ruithe et al., 2016). Maintaining consistency is crucial to integrating data, sharing data, and fostering collaboration among various groups within an organization.

1.8.1. Data Governance Frameworks and Best Practices

Organizations may utilize established procedures and best practices for executing effective data governance, which offers an organized method for managing data assets. The Data Management Body of Knowledge (DMBOK) is a framework that provides a thorough and all-encompassing overview of data management practices, which includes data governance. The DMBOK offers a structured approach to crafting a data governance plan, creating data governance committees, and outlining the duties and obligations of personnel within the enterprise.

Additionally, prominent data governance frameworks comprise the Data Governance Institute's Framework, which prioritizes the harmonization of data governance with commercial objectives, and the Control Objectives for Information and Related Technologies (COBIT), which accentuates data governance's control and security facets (Abraham et al., 2019). The aforementioned frameworks furnish a structured plan for entities to formulate their data governance tactics by drawing upon the most effective methodologies observed in the relevant field.

Apart from utilizing frameworks, various optimal methods can augment the efficiency of data governance endeavors. Organizations are recommended to commence their data governance efforts by formulating comprehensive data governance rules that delineate the goals, range, and fundamental tenets of data governance inside the organization. It is imperative that senior management effectively communicate and endorse this policy to secure organizational buy-in and support.

A pivotal element involves the establishment of a data governance council or committee comprising delegates from diverse business units and IT departments. The council operates as the authoritative entity accountable for formulating determinations pertaining to regulations, norms, and processes concerning data governance (Weber et al., 2009; Gudivada et al., 2017). It is recommended that the council convene on a regular basis to assess and revise the data governance structure in accordance with changing business demands and regulatory stipulations.

In addition, it is recommended that organizations allocate resources toward acquiring data governance technologies and tools that evaluate data quality, metadata management, and data lineage tracking. These instruments facilitate the monitoring of data quality, the identification of data anomalies, and the tracing of data flow across diverse systems for organizations.

1.8.2. Data Quality Management and Data Cleansing Techniques

Data quality management is an essential aspect of sophisticated data management systems, as it guarantees the precision, uniformity, entirety, and dependability of data. Inadequate data quality can result in significant ramifications, such as erroneous conclusions, inaccurate judgments, and jeopardized organizational results (Fürber & Hepp, 2011). Consequently, organizations must utilize data cleansing methodologies to detect and resolve any data quality concerns.

Remember

Implement regular data quality management practices, including data cleansing techniques such as deduplication and normalization, to ensure the accuracy and .reliability of your data

The process of data cleansing encompasses a sequence of procedures that are designed to enhance the quality of data by rectifying or eliminating data that is erroneous, incongruous, or replicated. Data profiling is a key method employed in the process of data cleansing. This technique involves a thorough examination and evaluation of data to detect any irregularities, deviations, or incongruities that may be present. Various data profiling techniques, such as statistical analysis, matching patterns, and rule-based validation, can be employed to detect possible data quality concerns.

Upon identification of data quality issues, organizations can utilize diverse data cleansing methods to resolve them. The employed methodologies may encompass the process of data standardization, which involves converting data into a standard form or representation, and data approval, which guarantees that data conforms to predetermined business regulations and limitations (Ridzuan & Zainon, 2019; Yu, 2014). The data cleansing process may encompass data enrichment, which pertains to the supplementation or enhancement of incomplete or deficient data through external data sources.

Additionally, entities can employ data deduplication methodologies to detect and remove redundant entries in their data repositories. The procedure entails comparing data attributes, such as names, addresses, or unique identifiers, to detect and consolidate or eliminate duplicate entries. This results in a reduction of redundancy and an enhancement of data quality.

1.8.3. Data Privacy and Compliance Considerations

The paramount significance of data privacy and compliance factors is evident within sophisticated data management systems (Capps, 2020). Organizations are obligated to comply with diverse legal and regulatory obligations, which may include the GDPR, Health Insurance Portability and Accountability Act (HIPAA), or industry-specific regulations, contingent upon the type of data they manage.

To guarantee adherence to data privacy regulations, it is recommended that organizations employ data masking or anonymization methodologies, particularly when disseminating data for the purposes of testing, development, or analysis. Data masking entails substituting sensitive or personally identifiable information (PII) with fabricated but plausible data, thereby safeguarding individuals' confidentiality while preserving the data's usefulness (Determann, 2017; Voss, 2016).

Furthermore, it is recommended that organizations implement controls on access and user permissions to limit data access according to specific roles and responsibilities. The implementation of access controls guarantees that solely authorized personnel can gain entry to and alter confidential information, thereby diminishing the likelihood of data breaches or unsanctioned data manipulation.

In addition, organizations must establish policies for data retention and disposal to manage the data lifecycle and adhere to legal obligations effectively. Policies must be established to delineate the temporal parameters for data retention and the protocols for the secure disposal of data that has exceeded its utility (Baloyi & Kotzé, 2020). It is imperative to utilize appropriate methods for disposing of data, such as secure data erasure or data shredding, to avert the possibility of information being compromised or unauthorized access.

Regularly reviewing and updating data governance procedures and standards is imperative for organizations to ensure alignment with developing privacy and compliance regulations. Adherence to privacy laws is not just legal liabilities but also bolsters consumer confidence and fortifies the entity's standing in question.

To summarize, the incorporation of data governance and data quality is imperative for sophisticated data management systems. Implementing policies, frameworks, and practices is crucial in establishing effective data governance, which aims to ensure data reliability, consistency, and security. Organizations can enhance their data management procedures and extract the maximum value from their assets by adopting data governance structures and adhering to best practices (Pokorný, 2006). Implementing data quality management techniques such as data profiling, data purification, and data deduplication is imperative in guaranteeing the precision and dependability of data. In addition, organizations must give precedence to data privacy and compliance factors, data masking techniques, access controls, and policies for data disposal. By adopting these guidelines and procedures, organizations can attain resilient and effective sophisticated systems for data management that facilitate informed decision-making and yield concrete business results.

1.9. EMERGING TRENDS IN ADVANCED DATA MANAGEMENT

In the swiftly developing domain of data management, organizations must remain abreast of emerging trends to proficiently manage the constantly expanding volume and intricacy of data (Casado & Younas, 2015; Shamim e al., 2019). The present chapter comprehensively analyzes several noteworthy new developments in sophisticated data management. The present study will explore the effects of big data, the IoT and sensor data management, blockchain technology, as well as distributed ledger databases, and the significance of AI and machine learning in data management.

1.9.1. Big Data and Its Impact on Data Management

1. **Definition and Characteristics of Big Data:** The term "big data" pertains to notably voluminous and heterogeneous datasets, rendering them challenging to handle and scrutinize through conventional data processing methods. The triad of volume, velocity, and variety distinguishes the phenomenon (Chaudhuri, 2012). The concept of big data encompasses three key dimensions: volume, velocity, and variety. Volume pertains to the vast quantity of data produced, velocity denotes the rapid pace at which data develops and necessitates handling, and variety denotes data's heterogeneous origins and structures.
2. **Challenges and Opportunities of Big Data:** Big data management encounters noteworthy challenges due to the vast amount and rapid rate of data accumulation, which also presents opportunities for data analysis. Conventional database management systems encounter difficulties in effectively managing extensive datasets. Nevertheless, the utilization of big data also offers significant prospects for enterprises (Siddiqa et al., 2016; Diène et al., 2020). Through the process of analyzing and gaining insights from large sets of data, enterprises can attain a competitive edge, enhance their decision-making capabilities, elevate customer experiences, and find novel sources of revenue.
3. **Technologies and Techniques for Big Data Management:** The present study focuses on the technologies and techniques employed for the management of big data. In order to proficiently handle large volumes of data, entities utilize sophisticated data management methodologies and technologies. One example of the distributed file system is the Hadoop Distributed File System (HDFS), which facilitates the distributed storage and processing of large-scale data across multiple clusters of low-cost hardware. Furthermore, parallel processing frameworks such as Apache Spark and Apache Flink offer advanced processing capabilities

with high performance for the purpose of big data analytics (Abu-Elkheir et al., 2013). Data lakes provide a scalable and adaptable solution for storing and analyzing large volumes of unstructured and unprocessed data. Sophisticated analytical methods, including data mining, modeling with predictions, and natural language processing, are employed to extract meaningful patterns and insights from large datasets.

1.9.2. Internet of Things (IoT) and Sensor Data Management

1. **Introduction to the Internet of Things (IoT):** It is a network of interconnected devices equipped with sensors, actuators, and connectivity abilities (Vongsingthong & Smanchat, 2015; Xiong et al., 2015). This technology has gained significant attention in recent years and has become an area of interest for researchers and practitioners alike. Autonomous devices can gather and share data independently, resulting in a significant surge of sensor data in real time. The spectrum of IoT applications encompasses diverse fields, including but not limited to intelligent domiciles and wearable technology, industrial process automation, urban infrastructure optimization, and medical care.
2. **Challenges of Sensor Data Management:** The management of sensor data poses various challenges. The exponential growth of the IoT devices and the uninterrupted flow of sensor-generated data pose distinctive obstacles to effective data management (Ali & Abu-Elkheir, 2012). The effective management of IoT data necessitates using scalable and distributed frameworks that can accommodate the substantial amount, speed, and variety of data produced. Furthermore, sensor data management necessitates crucial considerations such as data quality, integration, and real-time processing.
3. **Technologies and Techniques for Sensor Data Management:** Organizations utilize diverse technologies and techniques to proficiently manage sensor data in the realm of sensor data management (Cruz Huacarpuma et al., 2017; Ajao et al., 2019). The implementation of edge computing facilitates the computation and examination of data at the periphery of the network, in closer proximity to the origin of data production. This results in a decrease in both latency and bandwidth demands. Stream

processing frameworks such as Apache Kafka and Apache Flink enable the instantaneous analysis and processing of streaming sensor data. Time-series databases are a specialized type of database that is optimized for the storage and retrieval of time-stamped data (Kuo et al., 2011). These databases are particularly well-suited for handling data generated by IoT sensors, as they offer efficient and effective means of storage and retrieval.

1.9.3. Blockchain Technology and Distributed Ledger Databases

1. **Introduction to Blockchain Technology:** The present discourse aims to provide an overview of Blockchain Technology. Initially developed as the fundamental technology for digital currencies such as Bitcoin, Blockchain has garnered significant interest for its application in data management (Zachariadis et al., 2019; Mainelli & Smith, 2015). A distributed ledger is a technology that enables secure and tamper-proof recording of transactions or data throughout various parties. Utilizing blockchain technology provides benefits such as increased transparency, immutability, and distributed consensus.
2. **Applications of Blockchain in Data Management:** Organizations are currently investigating the potential of the blockchain system for diverse applications beyond the realm of cryptocurrencies, particularly in the field of data management. Implementing block chain technology in data management can potentially augment the levels of security, privacy, and data integrity (Swan, 2017). The technology facilitates secure and transparent data sharing among multiple entities, obviating the necessity for intermediaries and mitigating the likelihood of data tampering. Blockchain technology has the potential to be applied in various domains, such as managing supply chains, healthcare documents, money transfers, and digital identity management.
3. **Distributed Ledger Databases:** It have been influenced by blockchain technology and offer a distributed and copied storage mechanism for data. This mechanism guarantees the availability and consistency of data. The aforementioned databases are designed to ensure fault tolerance and endurance by storing several copies of the data throughout numerous nodes (Janssen et al.,

2020; Wang & Siau, 2019). Distributed ledger databases present several benefits, including decentralized control, enhanced data integrity, and audibility, rendering them appropriate for numerous parties requiring access to a communal database.

1.9.4. Artificial Intelligence (AI) and Machine Learning in Data Management

1. **Introduction to Artificial Intelligence (AI) and Machine Learning (ML):** The utilization of AI and ML methodologies has brought about a significant transformation in data management through the automation and enhancement of diverse tasks. AI pertains to creating systems that can replicate human intelligence, whereas ML centers on developing algorithms that empower computers to acquire knowledge from data and generate forecasts or judgments without the need for explicit programming (Lagaros & Plevris, 2022).
2. **AI and ML in Data Integration and Cleansing:** The utilization of AI and ML algorithms can aid in the facilitation of data integration and cleaning procedures. Automated discovery of connections among data elements, identification of duplicate records, and resolution of data inconsistencies can be achieved. Machine learning models can acquire knowledge from past data to enhance the evaluation of data quality and enhance the precision of data cleansing.
3. **AI and ML in Data Analysis and Decision Making**: Integrating AI and ML techniques in data analysis has facilitated organizations in discovering significant insights, anticipating future outcomes, and formulating decisions based on data (Sircar et al.,2021). Machine learning algorithms possess the capability to scrutinize extensive datasets for the purpose of recognizing patterns, detecting anomalies, and producing predictive models. AI driven systems can offer suggestions by analyzing data, thereby assisting organizations in enhancing their decision-making procedures (Prudius et al., 2019; Li et al., 2021).

Activity 1.1

You are working as a data architect for a large e-commerce company. Design a comprehensive data management strategy that incorporates both relational and NoSQL databases. Outline the key considerations you would consider when deciding which type of database to use for different data sets and scenarios. Additionally, discuss the challenges and benefits of using cloud databases for storing .and managing e-commerce data

1.10. SUMMARY

The chapter under consideration sheds light on the current developments in the realm of sophisticated data management, which signify a continuous evolution. Organizations' utilization and management of data are being transformed by various technological advancements such as big data, IoT and sensor data management, blockchain technology, and AI and machine learning. Incorporating these trends necessitates a confluence of technological progress, proficient personnel, and a deliberate methodology for data administration. Organizations that implement these new developments effectively will attain a competitive advantage by proficiently utilizing data to facilitate informed choices, innovation, and expansion. If organizations are to prosper in the data-driven future, they need to remain knowledgeable and flexible in response to the ever-changing landscape of the field.

REVIEW QUESTIONS

1. Explain the evolution of data management systems and the challenges they address.
2. Define advanced data management and its scope in modern organizations.
3. Discuss the benefits and advantages of advanced data management systems.
4. Describe the key components and concepts of relational databases.
5. What is data modeling, and how is it applied in relational databases?
6. Explain the role of SQL and advanced querying techniques in relational databases.
7. Discuss the types and categories of NoSQL databases.

8. What are the considerations for data modeling and design in NoSQL databases?
9. How do cloud computing and cloud databases impact data management?
10. Explain the concepts of data distribution, replication, and consistency in distributed databases.

REFERENCES

1. Abadi, D. J., (2009). Data management in the cloud: Limitations and opportunities. *IEEE Data Eng. Bull.*, *32*(1), 3–12.
2. Abdelhafiz, B. M., & Elhadef, M., (2021). Sharding database for fault tolerance and scalability of data. In: *2021 2nd International Conference on Computation, Automation and Knowledge Management (ICCAKM)* (pp. 17–24).
3. Abraham, R., Schneider, J., & Vom, B. J., (2019). Data governance: A conceptual framework, structured review, and research agenda. *International Journal of Information Management*, *49*, 424–438.
4. Abu-Elkheir, M., Hayajneh, M., & Ali, N. A., (2013). Data management for the internet of things: Design primitives and solution. *Sensors*, *13*(11), 15582–15612.
5. Agrawal, D., Das, S., & El Abbadi, A., (2011). Big data and cloud computing: Current state and future opportunities. In: *Proceedings of the 14th International Conference on Extending Database Technology* (Vol. 1, pp. 530–533).
6. Ajao, L. A., Agajo, J., Adedokun, E. A., & Karngong, L. (2019). Crypto hash algorithm-based blockchain technology for managing decentralized ledger database in oil and gas industry. *J*, *2*(3), 300–325
7. Al Shehri, W., (2013). Cloud database database as a service. *International Journal of Database Management Systems*, *5*(2), 1.
8. Alavi, M., & Leidner, D., (1999). Knowledge management systems: Issues, challenges, and benefits. *Communications of the Association for Information Systems*, *1*(1), 7.
9. Ali, N. A., & Abu-Elkheir, M., (2012). Data management for the internet of things: Green directions. In: *2012 IEEE Globecom Workshops, 1* (pp. 386–390).
10. Al-Ruithe, M., Benkhelifa, E., & Hameed, K., (2016). A conceptual framework for designing data governance for cloud computing. *Procedia Computer Science*, *94*, 160–167.
11. Anandhi, R., & Chitra, K., (2012). A challenge in improving the consistency of transactions in cloud databases-scalability. *International Journal of Computer Applications*, *52*(2), 12–14.
12. Atzeni, P., Bugiotti, F., Cabibbo, L., & Torlone, R., (2020). Data modeling in the NoSQL world. *Computer Standards & Interfaces*, *67*, 103149.

13. Bagui, S., & Nguyen, L. T., (2015). Database sharding: To provide fault tolerance and scalability of big data on the cloud. *International Journal of Cloud Applications and Computing (IJCAC)*, *5*(2), 36–52.
14. Baloyi, N., & Kotzé, P., (2020). Data privacy compliance benefits for organizations–A cyber-physical systems and internet of things study. In: *Information and Cyber Security: 18th International Conference, ISSA 2019, Johannesburg, South Africa, August 15, 2019, Proceedings 18* (pp. 158–172).
15. Bansal, S. K., & Kagemann, S., (2015). Integrating big data: A semantic extract-transform-load framework. *Computer*, *48*(3), 42–50.
16. Barton, D., & Court, D., (2012). Making advanced analytics work for you. *Harvard Business Review*, *90*(10), 78–83.
17. Begg, C., & Caira, T., (2012). Exploring the SME quandary: Data governance in practice in the small to mediumsized enterprise sector. *Electronic Journal of Information Systems Evaluation*, *15*(1), 3–13.
18. Berger, M. L., & Doban, V., (2014). Big data, advanced analytics, and the future of comparative effectiveness research. *Journal of Comparative Effectiveness Research*, *3*(2), 167–176.
19. Bhushan, A., Rastogi, P., Sharma, H. K., & Ahmed, M. E., (2016). I/O and memory management: Two keys for tuning RDBMS. In: *2016 2nd International Conference on Next Generation Computing Technologies, 1* (pp. 208–214). IEEE.
20. Bordoloi, S., & Kalita, B., (2013). Designing graph database models from existing relational databases. *International Journal of Computer Applications*, *74*(1).
21. Bose, I., & Mahapatra, R. K., (2001). Business data mining—A machine learning perspective. *Information & Management*, *39*(3), 211–225.
22. Breitbart, Y., & Korth, H. F., (1999). Replication and consistency in a distributed environment. *Journal of Computer and System Sciences*, *59*(1), 29–69.
23. Budgen, D., Rigby, M., Brereton, P., & Turner, M., (2007). A data integration broker for healthcare systems. *Computer*, *40*(4), 34–41.
24. Bugiotti, F., Cabibbo, L., Atzeni, P., & Torlone, R., (2014). Database design for NoSQL systems. In: *Conceptual Modeling: 33rd International Conference, ER 2014, Atlanta, GA, USA, October 27–29, 2014. Proceedings, 33*(pp. 223–231).

25. Capps, M. A., (2020). Determann's field guide to data privacy law: International corporate compliance. Lothar Determann. Cheltenham, UK, Northampton, MA: Edward Elgar Publishing, 2020, Pp. v, 212. ISBN: 9781789906202, US $89.95. *International Journal of Legal Information*, *48*(2), 86–87.
26. Casado, R., & Younas, M., (2015). Emerging trends and technologies in big data processing. *Concurrency and Computation: Practice and Experience*, *27*(8), 2078–2091.
27. Chakraborty, J., Padki, A., & Bansal, S. K., (2017). Semantic ETL—State-of-the-art and open research challenges. In: *2017 IEEE 11th International Conference on Semantic Computing (ICSC), 1* (pp. 413–418).
28. Chaudhuri, S., (2012). What next? A half-dozen data management research goals for big data and the cloud. In: *Proceedings of the 31st ACM SIGMOD-SIGACT-SIGAI Symposium on Principles of Database Systems* (pp. 1–4).
29. Chen, I. M. A., Chu, K., Palaniappan, K., Ratner, A., Huang, J., Huntemann, M., & Kyrpides, N. C., (2021). The IMG/M data management and analysis system v. 6.0: New tools and advanced capabilities. *Nucleic Acids Research*, *49*(D1), D751–D763.
30. Chen, J. K., & Lee, W. Z., (2019). An introduction of NoSQL databases based on their categories and application industries. *Algorithms*, *12*(5), 106.
31. Chidwick, J. M., (1996). Seismic data management systems: Issues and advantages. In: *SEG Technical Program Expanded Abstracts 1996* (pp. 993–996.)
32. Chung, C. W., (1990). DATAPLEX: An access to heterogeneous distributed databases. *Communications of the ACM*, *33*(1), 70–80.
33. Cruz, H. R., De Sousa, Jr. R. T., De Holanda, M. T., De Oliveira, A. R., García, V. L. J., & Kim, T. H., (2017). Distributed data service for data management in internet of things middleware. *Sensors*, *17*(5), 977.
34. Determann, L., (2017). Data privacy A to Z: International corporate compliance. In: *Determann's Field Guide to Data Privacy Law, 1* (pp. 106–187).
35. Di Vimercati, S. D. C., Foresti, S., Livraga, G., & Samarati, P., (2019). Data security and privacy in the cloud. In: *Mobile Multimedia/Image Processing, Security, and Applications 2019* (Vol. 10993, pp. 84–96).

36. Diène, B., Rodrigues, J. J., Diallo, O., Ndoye, E. H. M., & Korotaev, V. V., (2020). Data management techniques for internet of things. *Mechanical Systems and Signal Processing, 138*, 106564.
37. Dönderler, M. E., Şaykol, E., Arslan, U., Ulusoy, Ö., & Güdükbay, U., (2005). BilVideo: Design and implementation of a video database management system. *Multimedia Tools and Applications, 27*(1), 79–104.
38. Ehwerhemuepha, L., Gasperino, G., Bischoff, N., Taraman, S., Chang, A., & Feaster, W., (2020). HealtheDataLab–a cloud computing solution for data science and advanced analytics in healthcare with application to predicting multi-center pediatric readmissions. *BMC Medical Informatics and Decision Making, 20*, 1–12.
39. El-Seoud, S. A., El-Sofany, H. F., Abdelfattah, M., & Mohamed, R., (2017). Big data and cloud computing: Trends and challenges. *International Journal of Interactive Mobile Technologies, 11*(2).
40. Esposito, C., De Santis, A., Tortora, G., Chang, H., & Choo, K. K. R., (2018). Blockchain: A panacea for healthcare cloud-based data security and privacy? *IEEE Cloud Computing, 5*(1), 31–37.
41. Fry, J. P., & Sibley, E. H., (1976). Evolution of data-base management systems. *ACM Computing Surveys (CSUR), 8*(1), 7–42.
42. Fürber, C., & Hepp, M., (2011). Towards a vocabulary for data quality management in semantic web architectures. In: *Proceedings of the 1st International Workshop on Linked Web Data Management, 1* (pp. 1–8).
43. Giakoumakis, L., & Galindo-Legaria, C. A., (2008). Testing SQL server's query optimizer: Challenges, techniques, and experiences. *IEEE Data Eng. Bull., 31*(1), 36–43.
44. Gligorijević, V., & Pržulj, N., (2015). Methods for biological data integration: Perspectives and challenges. *Journal of the Royal Society Interface, 12*(112), 20150571.
45. Goel, S., & Buyya, R., (2007). Data replication strategies in wide-area distributed systems. In: *Enterprise Service Computing: From Concept to Deployment, 1* (pp. 211–241).
46. Golshan, B., Halevy, A., Mihaila, G., & Tan, W. C., (2017). Data integration: After the teenage years. In: *Proceedings of the 36th ACM SIGMOD-SIGACT-SIGAI Symposium on Principles of Database Systems, 1* (pp. 101–106).
47. Grad, B., (2012). Relational database management systems: The

formative years [Guest editor's introduction]. *IEEE Annals of the History of Computing*, *34*(4), 7, 8.

48. Grolinger, K., Higashino, W. A., Tiwari, A., & Capretz, M. A., (2013). Data management in cloud environments: NoSQL and NewSQL data stores. *Journal of Cloud Computing: Advances, Systems and Applications*, *2*, 1–24.
49. Gudivada, V., Apon, A., & Ding, J., (2017). Data quality considerations for big data and machine learning: Going beyond data cleaning and transformations. *International Journal on Advances in Software*, *10*(1), 1–20.
50. Gunawan, R., Rahmatulloh, A., & Darmawan, I., (2019). Performance evaluation of query response time in the document stored NoSQL database. In: *2019 16th International Conference on Quality in Research (QIR): International Symposium on Electrical and Computer Engineering, 1* (pp. 1–6).
51. Hayn, D., Veeranki, S., Kropf, M., Eggerth, A., Kreiner, K., Kramer, D., & Schreier, G., (2018). Predictive analytics for data driven decision support in health and care. *IT-Information Technology*, *60*(4), 183–194.
52. Holden, E. P., Kang, J. W., Anderson, G. R., & Bills, D. P., (2011). Databases in the cloud: A status report. In: *Proceedings of the 2011 Conference on Information Technology Education, 1* (pp. 171–176).
53. Hurson, A. R., Pakzad, S. H., & Cheng, J. B., (1993). Object-oriented database management systems: Evolution and performance issues. *Computer*, *26*(2), 48–58.
54. Huttunen, J., Jauhiainen, J., Lehti, L., Nylund, A., Martikainen, M., & Lehner, O. M., (2019). Big data, cloud computing and data science applications in finance and accounting. *ACRN Journal of Finance and Risk Perspectives*, *8*, 16–30.
55. Imam, A. A., Basri, S., Ahmad, R., Watada, J., & González-Aparicio, M. T., (2018). Automatic schema suggestion model for NoSQL document-stores databases. *Journal of Big Data*, *5*(1), 1–17.
56. Janssen, M., Brous, P., Estevez, E., Barbosa, L. S., & Janowski, T., (2020). Data governance: Organizing data for trustworthy artificial intelligence. *Government Information Quarterly*, *37*(3), 101493.
57. Jaradat, M., Jarrah, M., Bousselham, A., Jararweh, Y., & Al-Ayyoub, M., (2015). The internet of energy: Smart sensor networks and big data management for smart grid. *Procedia Computer Science*, *56*, 592–597.

58. Kanwar, R., Trivedi, P., & Singh, K., (2013). NoSQL, a solution for distributed database management system. *International Journal of Computer Applications*, *67*(2).
59. Karim, A., Siddiqa, A., Safdar, Z., Razzaq, M., Gillani, S. A., Tahir, H., & Imran, M., (2020). Big data management in participatory sensing: Issues, trends, and future directions. *Future Generation Computer Systems*, *107*, 942–955.
60. Kemper, A., Kossmann, D., & Zeller, B., (1999). Performance tuning for SAP R/3. *IEEE Data Eng. Bull.*, *22*(2), 32–39.
61. Khatri, V., & Brown, C. V., (2010). Designing data governance. *Communications of the ACM*, *53*(1), 148–152.
62. Koltay, T., (2016). Data governance, data literacy and the management of data quality. *IFLA Journal*, *42*(4), 303–312.
63. Kong, W., Lei, Y., & Ma, J., (2018). Data security and privacy information challenges in cloud computing. *International Journal of Computational Science and Engineering*, *16*(3), 215–218.
64. Kuo, T. T., Kim, H. E., & Ohno-Machado, L., (2017). Blockchain distributed ledger technologies for biomedical and health care applications. *Journal of the American Medical Informatics Association*, *24*(6), 1211–1220.
65. Lagaros, N. D., & Plevris, V., (2022). Artificial intelligence (AI) applied in civil engineering. *Applied Sciences*, *12*(15), 7595.
66. Leventhal, B., (2010). An introduction to data mining and other techniques for advanced analytics. *Journal of Direct, Data and Digital Marketing Practice*, *12*, 137–153.
67. Li, M., Gao, J., Zhou, C., Shen, X. S., & Zhuang, W., (2021). Slicing-based artificial intelligence service provisioning on the network edge: Balancing AI service performance and resource consumption of data management. *IEEE Vehicular Technology Magazine*, *16*(4), 16–26.
68. Linthicum, D. S., (2017). Cloud computing changes data integration forever: What's needed right now. *IEEE Cloud Computing*, *4*(3), 50–53.
69. Lucas-Estañ, M. C., Sepulcre, M., Raptis, T. P., Passarella, A., & Conti, M., (2018). Emerging trends in hybrid wireless communication and data management for the industry 4.0. *Electronics*, *7*(12), 400.
70. Mainelli, M., & Smith, M., (2015). Sharing ledgers for sharing economies: An exploration of mutual distributed ledgers (aka

blockchain technology). *Journal of Financial Perspectives*, *3*(3).

71. Malecha, G., Morrisett, G., Shinnar, A., & Wisnesky, R., (2010). Toward a verified relational database management system. In: *Proceedings of the 37th Annual ACM Sigplan-Sigact Symposium on Principles of Programming Languages* (pp. 237–248).
72. Malliga, P., (2012). Database services for cloud computing–an overview. *Database*, *2*(3).
73. Mao, R., Xu, H., Wu, W., Li, J., Li, Y., & Lu, M., (2015). Overcoming the challenge of variety: Big data abstraction, the next evolution of data management for AAL communication systems. *IEEE Communications Magazine*, *53*(1), 42–47.
74. Milani, B. A., & Navimipour, N. J., (2016). A comprehensive review of the data replication techniques in the cloud environments: Major trends and future directions. *Journal of Network and Computer Applications*, *64*, 229–238.
75. Milosevic, Z., Chen, W., Berry, A., Rabhi, F. A., Buyya, R., Calheiros, R. N., & Dastjerdi, A. V., (2016). Real-time analytics. *Big Data: Principles and Paradigms*, (2016), 39–61.
76. Mitchell, T. M., (1999). Machine learning and data mining. *Communications of the ACM*, *42*(11), 30–36.
77. Nargesian, F., Asudeh, A., & Jagadish, H. V., (2022). Responsible data integration: Next-generation challenges. In: *Proceedings of the 2022 International Conference on Management of Data* (pp. 2458–2464).
78. Nargesian, F., Asudeh, A., & Jagadish, H. V., (2023). Next-generation challenges of responsible data integration. In: *Proceedings of the Sixteenth ACM International Conference on Web Search and Data Mining* (pp. 1256–1259).
79. Oguntimilehin, A., & Ademola, E. O., (2014). A review of big data management, benefits, and challenges. *A Review of Big Data Management, Benefits and Challenges*, *5*(6), 1–7.
80. Podgorelec, V., & Kuhar, S., (2011). Taking advantage of education data: Advanced data analysis and reporting in virtual learning environments. *Elektronika ir Elektrotechnika*, *114*(8), 111–116.
81. Pokorný, J., (2006). Database architectures: Current trends and their relationships to environmental data management. *Environmental Modeling & Software*, *21*(11), 1579–1586.
82. Prudius, A. A., Karpunin, A. A., & Vlasov, A. I., (2019). Analysis of

machine learning methods to improve efficiency of big data processing in industry 4.0. In: *Journal of Physics: Conference Series* (Vol. 1333, No. 3, p. 032065).

83. Rallapalli, S., Gondkar, R., & Ketavarapu, U. P. K., (2016). Impact of processing and analyzing healthcare big data on cloud computing environment by implementing Hadoop cluster. *Procedia Computer Science*, *85*, 16–22.
84. Ren, K., Thomson, A., & Abadi, D. J., (2014). An evaluation of the advantages and disadvantages of deterministic database systems. *Proceedings of the VLDB Endowment*, *7*(10), 821–832.
85. Ridzuan, F., & Zainon, W. M. N. W., (2019). A review on data cleansing methods for big data. *Procedia Computer Science*, *161*, 731–738.
86. Rodríguez-Mazahua, L., Rodríguez-Enríquez, C. A., Sánchez-Cervantes, J. L., Cervantes, J., García-Alcaraz, J. L., & Alor-Hernández, G., (2016). A general perspective of BIG DATA: Applications, tools, challenges, and trends. *The Journal of Supercomputing*, *72*, 3073–3113.
87. Rodzi, N. A. H. M., Othman, M. S., & Yusuf, L. M., (2015). Significance of data integration and ETL in business intelligence framework for higher education. In: *2015 International Conference on Science in Information Technology (ICSITech)* (pp. 181–186).
88. Rothnie, Jr. J. B., Bernstein, P. A., Fox, S., Goodman, N., Hammer, M., Landers, T. A., & Wong, E., (1980). Introduction to a system for distributed databases (SDD-1). *ACM Transactions on Database Systems (TODS)*, *5*(1), 1–17.
89. Saed, K. A., Aziz, N., Ramadhani, A. W., & Hassan, N. H., (2018). Data governance cloud security assessment at data center. In: *2018 4th International Conference on Computer and Information Sciences, 1* (pp. 1–4).
90. Samarati, P., (2014). Data security and privacy in the cloud. In: *Information Security Practice and Experience: 10th International Conference, ISPEC 2014, Fuzhou, China, May 5–8, 2014. Proceedings 10* (pp. 28–41).
91. Schadt, E. E., Linderman, M. D., Sorenson, J., Lee, L., & Nolan, G. P., (2010). Computational solutions to large-scale data management and analysis. *Nature Reviews Genetics*, *11*(9), 647–657.
92. Schmidt, A., Waas, F., Kersten, M., Carey, M. J., Manolescu, I., & Busse, R., (2002). XMark: A benchmark for XML data management.

In: *VLDB'02: Proceedings of the 28th International Conference on Very Large Databases, 1* (pp. 974–985). Morgan Kaufmann.

93. Seenivasan, D., (2023). Extract, transform, load (ETL) best practices. *International Journal of Computer Trends and Technology*, *71*(1), 40–44.
94. Shamim, S., Zeng, J., Shariq, S. M., & Khan, Z., (2019). Role of big data management in enhancing big data decision-making capability and quality among Chinese firms: A dynamic capabilities view. *Information & Management*, *56*(6), 103135.
95. Shannak, R., Masa'deh, R. E., Al-Zu'bi, Z., Obeidat, B., Alshurideh, M., & Altamony, H., (2012). A theoretical perspective on the relationship between knowledge management systems, customer knowledge management, and firm competitive advantage. *European Journal of Social Sciences*, *32*(4), 520–532.
96. Sharma, H. K., Nelson, S. C., & Singh, S. K., (2016). Tuning I/O subsystem: A key component in RDBMS performance tuning. *Database Systems Journal*, *7*(1).
97. Siddiqa, A., Hashem, I. A. T., Yaqoob, I., Marjani, M., Shamshirband, S., Gani, A., & Nasaruddin, F., (2016). A survey of big data management: Taxonomy and state-of-the-art. *Journal of Network and Computer Applications*, *71*, 151–166.
98. Sircar, A., Yadav, K., Rayavarapu, K., Bist, N., & Oza, H., (2021). Application of machine learning and artificial intelligence in oil and gas industry. *Petroleum Research*, *6*(4), 379–391.
99. Smith, B. L., Lewis, D. C., & Hammond, R., (2003). Design of archival traffic databases: Quantitative investigation into application of advanced data modeling concepts. *Transportation Research Record*, *1836*(1), 126–131.
100. Sprague, Jr. R. H., (1995). Electronic document management: Challenges and opportunities for information systems managers. *MIS Quarterly*, 1–29–49.
101. Sreemathy, J., Brindha, R., Nagalakshmi, M. S., Suvekha, N., Ragul, N. K., & Praveennandha, M., (2021). Overview of ETL tools and talend-data integration. In: *2021 7th International Conference on Advanced Computing and Communication Systems (ICACCS)* (Vol. 1, pp. 1650–1654).
102. Sreemathy, J., Durai, K. N., Priya, E. L., Deebika, R., Suganthi, K.,

& Aisshwarya, P. T., (2021). Data integration and ETL: A theoretical perspective. In: *2021 7th International Conference on Advanced Computing and Communication Systems (ICACCS)* (Vol. 1, pp. 1655–1660).

103. Sreemathy, J., Nisha, S., & RM, G. P., (2020). Data integration in ETL using TALEND. In: *2020 6th International Conference on Advanced Computing and Communication Systems (ICACCS)* (pp. 1444–1448).
104. Strauch, C., Sites, U. L. S., & Kriha, W., (2011). NoSQL databases. *Lecture Notes, Stuttgart Media University*, *20*(24), 79.
105. Swan, M., (2017). Anticipating the economic benefits of blockchain. *Technology Innovation Management Review*, *7*(10), 6–13.
106. Tauro, C. J., Aravindh, S., & Shreeharsha, A. B., (2012). Comparative study of the new generation, agile, scalable, high performance NOSQL databases. *International Journal of Computer Applications*, *48*(20), 1–4.
107. Teng, X., & Gong, Y., (2018). Research on application of machine learning in data mining. In: *IOP Conference Series: Materials Science and Engineering* (Vol. 392, No. 6, p. 062202).
108. Trope, R. L., & Power, E. M., (2005). Lessons in data governance: A survey of legal developments in data management, privacy, and security. *Bus. Law*, *61*, 471.
109. Tudorica, B. G., & Bucur, C., (2011). A comparison between several NoSQL databases with comments and notes. In: *2011 RoEduNet International Conference 10th Edition: Networking in Education and Research, 1* (pp. 1–5).
110. Urban, S. D., & Dietrich, S. W., (2001). Advanced database concepts for undergraduates: Experience with teaching a second course. In: *Proceedings of the Thirty-second SIGCSE Technical Symposium on Computer Science Education, 1* (pp. 357–361).
111. Vassiliadis, P., (2009). A survey of extract–transform–load technology. *International Journal of Data Warehousing and Mining (IJDWM)*, *5*(3), 1–27.
112. Vongsingthong, S., & Smanchat, S., (2015). A review of data management in internet of things. *Asia-Pacific Journal of Science and Technology*, *20*(2), 215–240.
113. Voss, W. G., (2016). European union data privacy law reform: General data protection regulation, privacy shield, and the right to delisting.

The Business Lawyer, *72*(1), 221–234.

114. Wang, H., Xu, Z., Fujita, H., & Liu, S., (2016). Towards felicitous decision making: An overview on challenges and trends of big data. *Information Sciences*, *367*, 747–765.
115. Wang, W., & Siau, K., (2019). Artificial intelligence, machine learning, automation, robotics, future of work and future of humanity: A review and research agenda. *Journal of Database Management (JDM)*, *30*(1), 61–79.
116. Weber, K., Otto, B., & Österle, H., (2009). One size does not fit all—A contingency approach to data governance. *Journal of Data and Information Quality (JDIQ)*, *1*(1), 1–27.
117. White, C., (2005). Data integration: Using ETL, EAI, and EII tools to create an integrated enterprise. *Business Intelligence Journal*, *10*(I).
118. Winkelmann, R., Harrington, J., & Jänsch, K., (2017). EMU-SDMS: Advanced speech database management and analysis in R. *Computer Speech & Language*, *45*, 392–410.
119. Woods, L., István, Z., & Alonso, G., (2014). Ibex: An intelligent storage engine with support for advanced SQL offloading. *Proceedings of the VLDB Endowment*, *7*(11), 963–974.
120. Xiong, Z., Zhang, Y., Luong, N. C., Niyato, D., Wang, P., & Guizani, N., (2020). The best of both worlds: A general architecture for data management in blockchain-enabled internet-of-things. *IEEE Network*, *34*(1), 166–173.
121. Yin, S., & Kaynak, O., (2015). Big data for modern industry: Challenges and trends [point of view]. *Proceedings of the IEEE*, *103*(2), 143–146.
122. Yu, C. T., & Chang, C. C., (1984). Distributed query processing. *ACM Computing Surveys (CSUR)*, *16*(4), 399–433.
123. Yu, W. E., (2014). Data privacy and big data-compliance issues and considerations. *ISACA J*, *3*, 27–31.
124. Yubin, G., Liankuan, Z., Fengren, L., & Ximing, L., (2013). A solution for privacy-preserving data manipulation and query on NoSQL database. *Journal of Computers*, *8*(6), 1427–1432.
125. Zachariadis, M., Hileman, G., & Scott, S. V., (2019). Governance and control in distributed ledgers: Understanding the challenges facing blockchain technology in financial services. *Information and Organization*, *29*(2), 105–117.
126. Zhang, C., Naughton, J., DeWitt, D., Luo, Q., & Lohman, G., (2001).

On supporting containment queries in relational database management systems. In: *Proceedings of the 2001 ACM SIGMOD International Conference on Management of Data* (pp. 425–436).

127. Zhang, J., Liu, Y., Zhou, K., Li, G., Xiao, Z., Cheng, B., & Li, Z., (2019). An end-to-end automatic cloud database tuning system using deep reinforcement learning. In: *Proceedings of the 2019 International Conference on Management of Data, 1* (pp. 415–432).
128. Zhou, Q., (2016). Research on heterogeneous data integration model of group enterprise based on cluster computing. *Cluster Computing, 19*(3), 1275–1282.

CHAPTER

2 Fundamentals of Data Management

CONTENT

2.1. INTRODUCTION

In contemporary times, proficient administration of data holds significant importance for the prosperity of enterprises spanning diverse sectors in technology. The implementation of data management guarantees the proper arrangement, availability, and protection of data, thereby empowering enterprises to arrive at informed judgments, extract valuable perceptions, and attain a competitive advantage. The present chapter provides an in-depth analysis of the core principles of handling information on SQL and NoSQL, along with the dynamic developments within the cloud and structures.

RDBMS are considered the fundamental basis of contemporary data handling. A structured methodology is employed for data storage, involving tables, rows, along with columns to arrange data systematically. Databases with relationships provide numerous benefits, such as ensuring accuracy and uniformity and offering robust query capabilities. This chapter delves into the fundamental principles of database systems, their defining attributes, and the model of relationships that serves as the foundation for their construction (Berry & Linoff, 2000).\

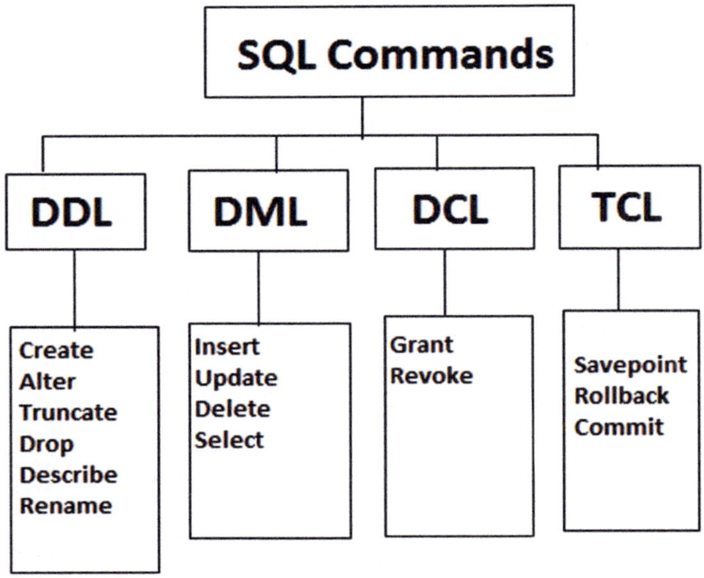

Figure 2.1. Schematic of types of SQL database.

Source: Technology crowds, Creative Commons License.

SQL is a universally accepted language that facilitates communication with relational database systems. The software allows individuals to establish and control information, execute inquiries, and oversee database organization. SQL is a programming language comprising two main components: the DDL, which stands for (Data Definition Language), and DML, which stands for DML statements. These components enable the efficient development, alterations, and retrieval of information. The chapter explores the fundamental constituents of SQL, encompassing data types, managers, gestures, and the syntax employed for prevalent operations (Figure 2.1).

The design of an information system is of paramount importance in the effective management of data. The design process is guided by standardization based on the relational framework. This approach helps to eliminate data anomalies and minimize redundancy. This chapter delves into the concept of normalization forms, namely 1NF, 2NF, and 3NF, and their crucial role in attaining a database design that is both streamlined and optimized. Furthermore, the material presents entity-relationship (ER) simulation, a methodology used for entities, connections, and characteristics in the conceptual design of databases (Koltsidas & Viglas, 2011).

Efficient databases rely heavily on optimizing queries and system tuning for optimal functionality. The optimization of queries involves the conversion of requests from users by considering various factors, including selectivity, searching, and join operations. The performance tuning process encompasses the refinement of database arrangement, the optimization of index utilization, and the implementation of caching processes. The chapter provides an in-depth analysis of the optimization of queries along with performance tuning techniques, presenting valuable perspectives and optimal methodologies for enhancing query speed of execution and overall system efficiency.

The development of NoSQL database systems as a replacement for conventional relational ones can be attributed to the development of big information and the necessity to manage extensive volumes of semi-structured and unstructured information. Non-relational databases, commonly known as NoSQL databases, provide advantages such as enhanced scalability, adaptability, and superior performance. This chapter delves into the underlying incentives over NoSQL databases, delineates their distinctive attributes, and surveys the various classifications of NoSQL in the form of databases, including document-oriented, key-value, vertical, and diagram (Sallans & Lake, 2014).

The chapter's concluding section emphasizes the significance of comprehending the basic principles of data leadership, such as ACID properties, cooperation control, separation levels, and preventing deadlocks. The concepts above hold significant importance in guaranteeing data credibility, uniformity, and effectiveness across interpersonal non-SQL database circumstances. Experts within data management might efficiently layout, optimize, and supervise databases by thoroughly understanding the fundamental principles. This enables them to fully exploit the immense value of their information assets within the present data-centric world (Provost & Fawcett, 2013) (Figure 2.2).

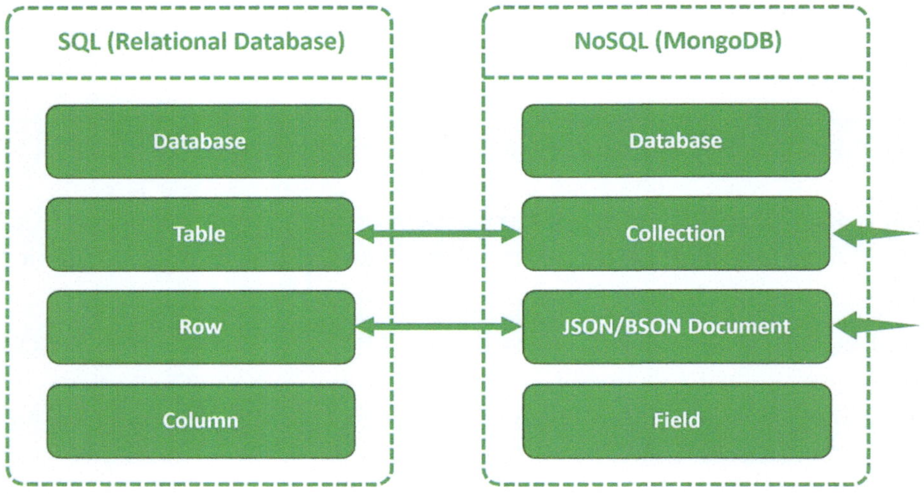

Figure 2.2. Illustration of difference between SQL and Non-SQL.

Source: Pragimtech, Creative Commons License.

2.2. RELATIONAL DATABASES

Databases with relationships play a fundamental role in data management within modern computer systems. Structured approaches are utilized for arranging and handling data, providing a dependable and effective means of storing, retrieving, and manipulating substantial amounts of data. The next part will examine the meaning and features of relational database systems, along with the basic ideas for the relational structure, which include row, column, and table structure (Yang, et al., 2009; Mishra & Eich, 1992; Buckles & Petry, 1982; Teorey, et al., 1986).

2.2.1. Definition of Relational Databases

A database with a relational structure is a structured assembly of interrelated data arranged in a tabular format, wherein data entities are stored in rows, and connections among those tables are created through common attributes. The term "relational" originates from depicting connections among entities by linking shared data attributes throughout tables. The relational structure facilitates streamlined and adaptable querying, the influence of data, and the implementation of security measures.

2.2.2. Characteristics of Relational Databases

Databases with relationships are widely utilized and popular due to several basic features that they possess:

1. **Tabular Structure:** Relational databases utilize a tabular structure consisting of columns and rows for storing data. Every table corresponds to a distinct entity or idea in the tangible world, while the columns symbolize the characteristics or qualities of that particular entity. The columns, documents, or tuples comprise the factual data situations or values.
2. **Data Independence:** Databases with relationships between the physical preservation of data and its logical depiction. The partitioning of data enables independence from data, wherein the mathematical framework of the relational database may be utilized for writing applications or queries without being impacted by modifications in its physical location or arrangement.
3. **Data Integrity:** Integrity rules are enforced by SQL databases to preserve the coherence and precision of the data. Integrity constraints, including main key and foreign key limitations, verify constraints, guarantee data validity, preserve connections, and fulfill specific conditions.
4. **Query Capability:** Structured Query Language, also called SQL, is a robust query language supported by relational database systems. SQL enables users to extract, modify, and merge data from single or multiple tables through declarative statements. The interpersonal databases' ability to query facilitates the effective recovery of data according to diverse criteria and lets intricate data analysis.

5. **ACID Properties:** Relational databases conform to the ACID (Durability, Atomicity, Uniformity isolate) principles to guarantee dependable and resilient transaction processing. The concept of atomicity ensures that a transaction is regarded as an unbreakable and single section of work. The maintenance of a valid state of the database both before and after every transaction is ensured by consistency. The implementation of isolation prevents interference between concurrent actions; strength assures that any changes dedicated will remain enduring and survive any potential system errors.

2.2.3. Relational Model: Tables, Rows, and Columns

The interpersonal model offers a structured look at arranging data inside an interpersonal database. It is centered on tables, rows, and posts, which create the basic developing hinders for the interpersonal database framework.

- **Tables:** Tables have a central organization within an interpersonal database. A set of interrelated data instances denotes the entities above or concepts. Every table possesses a distinctive name along with comprises rows along with columns. A corporate database could comprise distinct staff, divisions, and project tables.
- **Rows:** Rows, additionally referred to as documents or tuples, represent particular information instances within a piece of furniture. Every row in the table comprises data values related to the cells' characteristics or columns. For instance, in an employee surface, each row might indicate a staff, including data like staff ID, name, location, and earnings.
- **Columns:** Columns, referred to as characteristics and fields, specify the information's features or qualities in a given table. Every column possesses a distinct appellation and a particular data classification that ascertains what data it can accommodate. As an illustration, within a table of employees, the columns may comprise characteristics such as the employee's identification number, full name, and corresponding salary.

The relational framework facilitates connections between tables by utilizing common characteristics or keys. The relational file can aptly depict intricate interactions and facilitate streamlined data retrieval via joins and additional relational activities by establishing links between interrelated tables based on shared attributes.

To conclude, relational database systems offer a methodical and effective approach to data management. The relational framework, which utilizes tables of information, rows, and columns, is the fundamental basis of relational databases. This model facilitates the systematic and coherent organization, retrieval, and modification of data.

2.3. STRUCTURED QUERY LANGUAGE (SQL)

SQL represents a standardized and potent language facilitating interaction with relationship-based database management software (RDBMS). The tool offers extensive commands and constructs that enable data management and manipulation within databases. This section will comprehensively explore the diverse facets of SQL, encompassing its overview, significance in relational database management systems, data types, managers, expressions, and pivotal DDL and DML statements. Furthermore, we shall delve into rudimentary SQL queries. The four fundamental operations in relational databases are SELECT, DELETE, UPDATE, or INSERT.

Did You Know?

Domestic robot is a type of service robot, an autonomous robot that is primarily used .for household chores, but may also be used for education, entertainment or therapy

2.3.1. Importance of SQL in RDBMS

SQL is an explicit programming syntax developed to interact with RDBMS. It provides syntax and semantics, permitting users to interact with records. SQL is essential in managing information as it offers a uniform method for executing database operations throughout different systems vendors. The extensive utilization of this language has established it as the default choice for interacting with relational databases (Imieliński & Lipski, 1984).

SQL provides users diverse functionalities for manipulating databases, such as the ability to generate and alter database structures, retrieve, and query data, insert novel records, update pre-existing data, and remove records. The crucial skill set possessed by database managers, programmers, experts, and data researchers is attributed to the versatility and user-friendliness of the tool (Sahoo, et al., 2009).

2.3.2. SQL Data Types, Operators, and Expressions

SQL provides support for a range of data types that specify the nature of the data, which can be accommodated within database columns of data. Practical SQL types of information involve integers, floats, fractions, strings, time frames, and Booleans (Date, 1988). Comprehending the various data types at one's disposal and making informed decisions regarding their selection is paramount for optimizing data preservation and retrieval processes. This is because such decisions have a direct impact on the storage capacity, accuracy, and scope of values that can be effectively represented. Workers in SQL enable users to carry out operations upon data. SQL offers a range of operators, including arithmetic operators such as addition (+), subtraction (-), multiplication (*), and division (/), comparison operators such as equals (=), not equals (<>), greater than (>), less than (<), greater than or equal to (>=), and less than or equal to (<=), logical operators such as AND, OR, and NOT, as well as a string encoding operator (||), among others (Figures 2.3, 2.4 & 2.5).

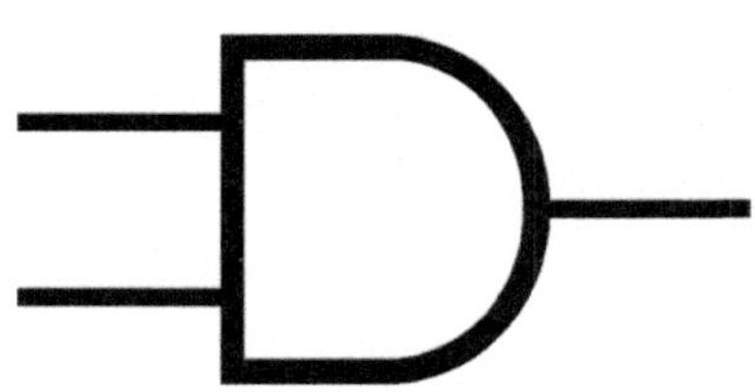

Figure 2.3. Sign of AND operator.

Source: Wikimedia, Creative Commons License.

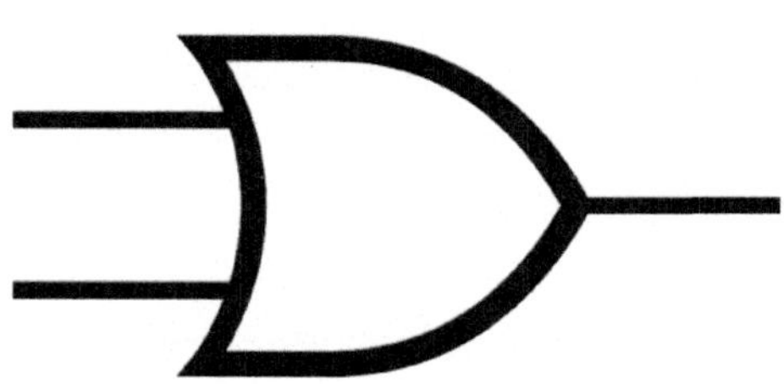

Figure 2.4. Sign of OR operator.

Source: Wikimedia, Creative Commons License.

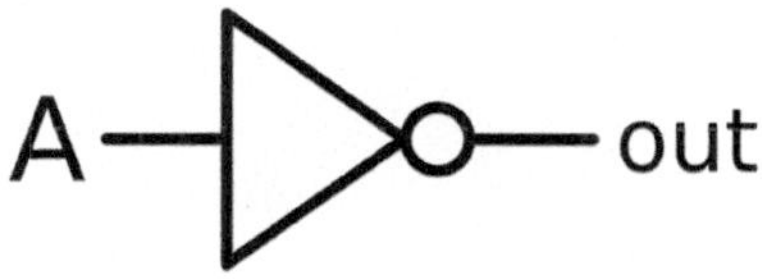

Figure 2.5. *Sign of NOT operator.*

Source: Wikimedia, Creative Commons License.

Using employees within SQL queries allows for the execution of calculations, contrasts, and mathematical operations by users (Rys, 2005; Armbrust, et al., 2015, Kulkarni & Michels, 2012).

SQL phrases operators, operations, and literals, along with column values, to generate a singular value. SQL phrases are commonly utilized in various sections of SQL queries, such as WHERE provisions and choose financial statements. The platform enables users to perform computations, manipulate data, and apply filters to outcomes based on predetermined criteria. SQL offers a diverse range of pre-defined functions that can be utilized in expressions, encompassing numerical, string, date, and assembling functions (Lakshmanan, et al., 1999; Melton, et al., 2002; Eisenberg, et al., 2002).

2.3.3. Data Definition Language (DDL) and Data Manipulation Language (DML) Statements

SQL is split into two main groups about statements: Data Definition, The use of language (DDL), and DML. Statements in DDL are utilized for defining and handling the structure of records and objects in them. The fundamental DDL assertions comprise CREATE, DROP, or ALTER. The CREATE statement generates novel objects such as databases, indexes, views, tables, and other related entities. The ALTER declaration enables users to alter the structure of previously created objects. The DROP declaration is a command to remove databases, tables of information, and additional objects (Amornchewin, 2018; Ramirez, et al., 1974).

In contrast, DML statements are employed to modify data stored within a database. The statements encompassed within this discussion's scope are SELECT, INSERT, UPDATE, and DELETE. The SELECT operator is the fundamental SQL statement utilized to retrieve data from any number of tables. The system enables users to extract specific columns, apply row-based filtering according to specified conditions, and execute diverse operations, including sorting and arrangement (Parsons, et al., 1974, Ramirez, 1973; Setiyadi, 2021).

The INSERT declaration is a SQL command for adding new data records into a table. This functionality allows users to efficiently input one or more rows containing information into designated columns within a given table. The UPDATE declaration is a SQL command to alter or modify pre-existing data within a table. The feature enables users to modify particular rows or columns by applying specific criteria. The DELETE declaration is a database operation that facilitates the removal of information from the table according to predetermined conditions (Tiffany, 2003; Kau & Tseng, 1994; Wells, 2001).

2.3.4. Basic SQL Queries: SELECT, INSERT, UPDATE, DELETE

The SELECT operator constitutes the fundamental element of SQL inquiries. The system enables users to extract information from singular or multiple tables by applying predetermined criteria. The SELECT statement provides diverse clauses, such as WHERE, GROUP BY, HAVING, ORDER BY, and JOINs. Utilizing such clauses facilitates data filtering, aggregation, sorting, and combination from various tables, producing significant result sets (Agrawal, et al., 2005; Chaudhuri & Narasayya, 1997).

The INSERT declaration is a SQL command that adds fresh records to a database table. The feature enables users to designate specific values for insertion or retrieval from alternative tables. Including information databases, whether a solitary record or a large quantity of data, relies upon the pivotal INSERT statement. The UPDATE declaration is a SQL command utilized to alter or adjust the values of the current information in a specified table. The feature lets users modify columns or rows contingent on predetermined conditions. This assertion is crucial in upholding data precision and effecting modifications to the extant records. The DELETE declaration is a SQL command utilized to eliminate information within a table that has been modified. The feature enables users to selectively remove rows by applying predetermined criteria. The DELETE operation holds significant importance in maintaining the confidentiality of data and eliminating redundant or outdated records about databases (Van, et al., 2007; Kearns, et al., 1997; Koonce, et al., 1998).

In summary, SQL seems a potent language that facilitates the management and manipulation of data within relational database systems. This section examines SQL's general outline and importance within RDBMS. During the session, the operators, syntax of SQL types of data, managers, and expressions were deliberated upon, as they serve as the fundamental building blocks for modifying data and evaluations. Furthermore, we discussed the fundamental DDL and DML statements that establish the database framework and enable data manipulation. The study encompassed an analysis of the primary SQL queries, namely SELECT, INSERT, UPDATE, and DELETE, that constitute the fundamental procedures involved in the data (Aboutorabi, et al., 2015; Truica, et al., 2013).

2.4. NORMALIZATION AND DATABASE DESIGN

2.4.1. Importance of Normalization in Database Design

The concept of normalization holds significant importance in the realm of database design because it strives to eradicate duplication and enhance the integrity of data. Normalization is a process that enhances the efficiency, flexibility, and maintainability of a database by arranging data into properly organized tables and minimizing data redundancy (Kennedy, 2000; Shoval & Even-Chaime, 1987). Minimizing data anomalies, including revision, inserting, and deletion deviations, ensures data is kept consistently and without redundancy (Figure 2.6).

Employee

employeeID	employeeName	managerID	sectorID
1	David D.	1	4
2	Eugene E.	1	3
3	George G.	2	2
4	Henry H.	2	1
5	Ingrid I.	2	4
6	James J.	3	1
7	Katy K.	3	4

Sector

sectorID	sectorName
1	Administration
2	Security
3	IT
4	Finance

Manager

managerID	managerName	area
1	Adam A.	East
2	Betty B.	West
3	Carl C.	North

Figure 2.6. Image of normalized database.

Source: Phoenix NAP, Creative Commons License.

The principal objective of normalizing is to attain data integrity, which guarantees the accuracy and consistency of data across the database. Normalization is a process that involves the removal of redundant data and the establishment of associations among entities. This process is aimed at minimizing the occurrence of contradictions along with contradictions. This aspect holds significant importance in environments that involve several individuals or applications that may concurrently access and alter the database (Ceri & Gottlob, 1986; Jahan & Edwards, 2015).

2.4.2. Functional Dependencies and Normalization Forms (1NF, 2NF, 3NF)

Dependencies on function are of utmost importance in the process of standardization. A table establishes a functional connection when a particular attribute's contents can ascertain another characteristic's appreciation. Comprehending functional dependencies facilitates the identification of primary keys, ascertaining the connections between organizations, and implementing normalizing rules.

The normalization process encompasses multiple normalization forms, with each subsequent form developing upon the preceding one.

1. According to the principle of the first usual form (1NF), it is necessary for every attribute present in a table to be atomic, which implies that it should consist solely of indivisible values. The process of eliminating duplicate groups and guaranteeing the uniqueness of every row within a table is achieved (Dongare, et al., 2011).
2. The additional normal shape (2NF) concept centers on removing limited dependencies. In database design, a partial dependence happens when an attribute is dependent solely on a portion of a primary key that is a composite. The second normal form (2NF) is achieved by breaking down a table through smaller ones and appropriately assigning characteristics to the following table. This process ensures that every attribute depends on the entire main key (Albarak, & Bahsoon, 2018; Dongare, et al., 2011)
3. The third standard form (3NF) concept pertains to eliminating temporary dependencies. In database design, a transitive dependency is said to arise when a property is functionally dependent on another property that does not constitute a constituent of the primary key. The third normal form (3NF)

can attain a greater degree of normalizing and minimize data redundancy by breaking down tables and removing dependencies (Chen, et al., 2007; Bahmani, et al., 2008; Demba, 2013).

2.4.3. Denormalization and Trade-offs

Although normalization is essential for preserving the confidentiality of information and minimizing redundant operations, there are still instances in which normalization might be suitable. Denormalization is a deliberate technique employed in database design to enhance performance by minimizing the number of joins necessary to retrieve data by introducing redundancy.

This process is commonly employed in scenarios with a high emphasis on read operations or when intricate queries must be carried out. Denormalization is a technique that involves storing redundant data in a database. This approach may assist in lowering the amount of table connects necessary to retrieve information, which in turn can enhance the speed of queries.

Nevertheless, denormalization entails certain trade-offs. The utilization of this approach may result in an escalation of data storage demands, the introduction of data inconsistencies in the event of inadequate update management, and the possible complication of data adaptations (Abyzov, 2023; Sanders, et al., 2001; Shin & Sanders, 2006). The decision of whether to renormalize necessitates meticulous contemplation of the demands of the program, the anticipated workload, and its performance ramifications. It is crucial to evaluate the potential advantages compared with the disadvantages and analyze the effect on the integrity of information and the complexity of upkeep (Helland, 2019).

2.4.4. Entity-Relationship (ER) Modeling and Database Design

The Entity-Relationship (ER) modeling process is widely utilized for conceptualizing and designing databases. The system is depicted visually by illustrating its constituent entities, deemed significant, and their interconnections. ER diagrams utilize diverse terminologies and symbols to show entities, attributes, connections, and cardinality.

Within ER modeling, a person or thing is used to signify a tangible or intangible object or notion within the real world. Examples of such entities may include a client as well as an order. Attributes refer to the inherent features or qualities of entities, including the nomenclature or chronological age of a patron. The concept of relationships pertains to how entities are interconnected, exemplified by the correlation between a patron and a purchase order (Dhabe, et al., 2010) (Figure 2.7).

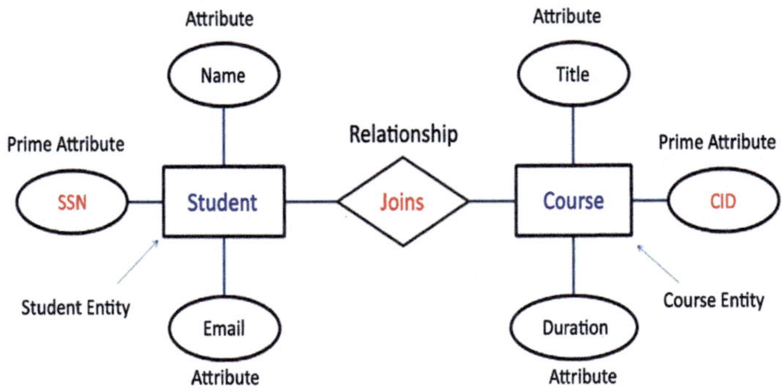

Figure 2.7. Illustration of entity-relationship (ER) modeling.

Source: LCSO, Creative Commons License.

ER modeling facilitates comprehension of the database's structure, recognition of organizations' characteristics, and visualization of the interconnections among entities. The ER model provides a fundamental basis for the subsequent logical design phase, wherein it is transformed into a database's structure. The resultant schema encompasses various components such as rows, posts, primary and secondary keys, and additional constraints that are imperative to the system's database design (Teorey, et al., 1986).

A comprehensive comprehension of the business demands, and the inter-entity relationships is imperative for proficient ER simulation. The process aids in verifying that the database architecture effectively reflects actual situations in the physical world and facilitates the intended tasks of the entire system (Hussain, et al., 2003).

Incorporating normalizing or database layouts is essential in constructing effective, resilient, and sustainable databases. Normalization is imperative in maintaining data integrity and minimizing redundancy through systematically arranging data into structured tables. Comprehending functional dependencies and implementing normalization forms such as the first normal form (1NF), second normal form (2NF), and third normal form (3NF) can facilitate the attainment of elevated levels of the normalization process. Denormalization might be employed as a strategic measure to enhance performance, albeit at the expense of introducing trade-offs. Moreover, ER modeling offers a graphical depiction of relationships, attributes, and entities and simplifies the creation of databases. Through the implementation of those concepts and methods, database designers can construct database architectures that are improved and scalable, thereby satisfying the demands of the software or system (Bock, 1997; Embley & Ling, 1999).

2.5. QUERY OPTIMIZATION AND PERFORMANCE TUNING

Optimization of queries and performance adjusting are essential components of handling information in any RDBMS. The procedures above are designed to enhance the efficacy and velocity of data retrieval procedures, guaranteeing peak performance even when dealing with extensive datasets. The forthcoming section will undertake an in-depth examination of various query optimization, methodologies, optimization algorithms, executing query strategies, indexed techniques (which include B-trees, encode indexes, along with a bitmap index), as well as performance enhancement factors, including searching, table partitioning, along with caching (Kamatkar, et al., 2018).

Improving queries in RDBMS entails examining and restructuring queries to reduce the computing power needed for their implementation. The aim is to produce an optimal plan for executing a query that minimizes the time required or maximizes the use of resources. RDBMS utilize advanced query optimization techniques that use statistical data, cost-based analysis, and heuristics to ascertain the optimal method for executing a given query. The creation of question execution strategies is a crucial element in optimizing queries. As mentioned earlier, the plans delineate the sequential procedures that the RDBMS will undertake to procure the data that has been solicited. The optimization process considers multiple variables, including but not limited to the dimensions of the data table, the number of indexes available, and the resources the system has at its disposal to determine the optimal plan of execution. The execution plans may be depicted through various formats, such as textual depictions, visualizations, or a series of operations. The execution plans of queries offer valuable insights into the strategies that the RDBMS will use in accessing and joining tables, implementing filters along with aggregations, and utilizing optimization methods such as indexes (Schiefer & Valentin, 1999).

The RDBMS utilizes algorithms for optimization to produce the most efficient execution plan. The methods under consideration assess various tactics and merging techniques while considering factors like selectivity, cardinality approximations, and expenses. The optimization process involves an analysis of various plans, accompanied by selecting the strategy via the lowest calculated cost, as determined by the chosen optimization algorithm. Exhaustive look, dynamic code, and heuristics-based methods are frequently employed in algorithms for optimization. The algorithms evaluate the compromises among various approaches for executing queries to ascertain the optimal plan with the highest efficiency. The process of indexing is crucial in enhancing the efficiency of query execution. Data recovery can be greatly sped up by building the right indexes. B-trees, hashing indexes, and image indexes are among the most prevalent indexing techniques in RDBMS. B-trees are frequently used for equality checks and range queries. By reducing the amount of disk I/O needed, they preserve a healthy tree structure, enabling effective search operations. B-trees are highly useful in cases when the data is constantly updated (Li, et al., 2010).

Hash indexes are appropriate for precise match searches. They quickly perform lookup tasks using hashing algorithms to assign keys to particular locations in storage. Hash indexes are particularly effective for read-intensive applications having low update frequency. Bitmap indexes are a valuable tool for columns with low cardinality. In such cases, the index stores a bitmap for every unique value. The utilization of indexes facilitates expedited recovery of records through bit manipulations, rendering them especially effective for decision-making systems and information warehouse management.

Performance optimization entails more than just the optimization of queries and indexing. Implementing supplementary methodologies, like table splitting and caching, may augment the system's overall efficiency. The process of table splitting entails the fragmentation of extensive tables through smaller, more controllable partitions, which are categorized based on a predetermined criterion, including a particular column or an assortment of values.

The partitioning technique facilitates the execution of queries in parallel and enhances the speed of data retrieval by diminishing the volume of data that necessitates scanning. The implementation of data pruning enables efficient data management and has the potential to notably enhance query speed, particularly when dealing with extensive datasets (Habimana, 2015). Caching is a highly effective method for optimizing efficiency. Storing frequently within memory, known as caching, enables subsequent queries to be served immediately from the cache, removing the requirement for costly disk input/output operations. The use of caching processes may be deployed at diverse tiers, such as query outcome caching, caching at the database level, or caching at the application level. The practice of storing the outcomes of frequently carried out queries, which enables subsequent deaths to be served via the cache, is known as query outcome caching. Database-level caching entails caching complete tables from databases or frequently obtained data structures within the memory. Implementing an application-level cache involves integrating it into the application logic, where it selectively caches particular data or computation outcomes to minimize the frequency of database access. To achieve optimal performance, it is imperative to consider various factors, including cache dimensions and uniformity, when devising appropriate caching tactics (Colley & Stanier, 2017).

To summarize, optimizing queries and adjusting for efficiency are crucial aspects in ensuring effective data management within (RDBMS. The enhancement of system efficiency can be achieved through the utilization of query optimization methods, efficient execution strategy generation, and the implementation of indexing strategies that include B-trees, encode indexes, and bitmaps indexes. Performance-tuning factors like searching, table division, and caching should also be considered. A designed database system is crucial for enhancing data retrieval speed, minimizing resource utilization, and enhancing users' overall experience. Professionals working with RDBMS must possess a comprehensive understanding of the methodologies and considerations for optimizing performance. This is essential to ensure efficient accessibility for end-users (Kemper, et al., 1999; Schnaitter, et al., 2006).

2.6. NOSQL DATABASES

2.6.1. Evolution and Motivations Behind NoSQL Databases

Non-relational databases, commonly called NoSQL databases, have become a viable substitute for conventional RDBMS like Oracle, MySQL, and SQL Server. The acronym "NoSQL" denotes "Not Only SQL," implying that these databases offer a departure from the inflexible organization and querying functionalities of SQL-based systems. The emergence of NoSQL databases may be attributed to the beginning of the 2000s, during which web-scale enterprises such as Facebook, Amazon, and Google encountered difficulties effectively handling their swiftly expanding data (Scherzinger, et al., 2006).

The principal impetus for the emergence of NoSQL databases proved the necessity to manage copious quantities of unstructured or partially structured data produced by web-based applications. RDBMS encountered difficulties in horizontal scaling and delivering the required efficiency for these workloads due to their inflexible schemas and queries being processed on joins. In contrast, NoSQL databases provided a data structure that was more adaptable and structures that were distributed, enabling seamless expansion to accommodate extensive amounts of data (Mohamed, et al., 2014; Hillenbrand, et al., 2020).

The pursuit of high levels of availability and tolerance for faults was another influential factor in the emergence of NoSQL databases. Despite potential hardware malfunctions or network partitions, web-based applications frequently require an uninterrupted operation. NoSQL databases have adopted eventual consistency examples and distributed reproduction strategies to guarantee the persistence and accessibility of data, even in unfavorable circumstances (Scherzinger & Sidortschuck, 2020).

2.6.2. Characteristics and Advantages of NoSQL Databases

NoSQL databases possess distinct features that differentiate them from conventional RDBMS. Comprehending these attributes is imperative to comprehend the advantages they provide:

1. **Flexible Data Model:** NoSQL databases facilitate schema-less architecture, enabling data to be maintained in a more dynamic and adaptable format in contrast to the inflexible tabular arrangement of a relational database Document-oriented files, such as those that utilize JSON along with BSON format, possess the capability to conform to evolving needs for data readily (Han, et al., 2011).
2. **Horizontal Scalability:** The architecture of database systems using NoSQL is intended to facilitate horizontal scaling, which involves data distribution across numerous servers or clusters. The ability to scale enables organizations to manage substantial information and meet traffic by incorporating additional ordinary servers (Moniruzzaman, et al., 2013).
3. **High Performance:** NoSQL databases may improve writing and reading, emphasizing versatility and scalability. Frequently,

denormalization, which caching processes and optimized preservation formats are utilized to augment the velocity of retrieving information and processing.

4. **Distributed Architecture:** No-SQL database systems are designed to function in distributed settings using distributed tables of identifiers and shared-nothing designs. The strategy mentioned above guarantees fault tolerance and elevated availability using data replication across numerous nodes, enabling local execution of operations (Nayak, et al., 2013).
5. **Support for Unstructured Data:** No-SQL databases are proficient in managing various data types, including feeds from social media, information from sensors, or log records. The pliability of their data models facilitates effortless preservation and retrieval of heterogeneous data categories, obviating the necessity for substantial schema alterations.
6. **Agile Development:** Utilizing NoSQL databases provides developers with increased flexibility and adaptability by facilitating swift prototyping and continual improvement. These tools facilitate rapid iterations and diminish the time to market for novel features or uses by eliminating the necessity for advanced schema design (Sharma & Dave, 2012).

2.6.3. Choose NoSQL Over RDBMS

Although databases built with NoSQL along with RDBMS each have respective merits and demerits, there exist particular circumstances where opting for NoSQL could prove to be more beneficial:

1. **Large-Scale Data:** NoSQL databases demonstrate their superiority when confronted with extensive data amounts surpassing a solitary server's abilities. Horizontal scaling enables data distribution across numerous nodes, thereby facilitating the processing thousands of gigabytes of information (Kunda & Phiri, 2017).
2. **Flexible Data Requirements:** Databases using NoSQL provides greater flexibility for data that undergoes frequent modifications or has an intricate structure that is not easily accommodated within a fixed tabular schema. Document-oriented records are particularly suitable for situations where the structure may undergo changes or vary throughout various entities.

3. **High Performance and Low Latency:** NoSQL databases are specifically engineered to facilitate rapid data retrieval with minimal delay and handle large volumes of data processing efficiently. No-SQL database systems can offer notable performance improvements over conventional RDBMS for applications requiring swift information retrieval or immediate analysis (Han, et al., 2011).
4. **Agile Development and Scalability:** NoSQL databases facilitate agile creation methods and can scale effortlessly to cope with expanding user populations or evolving demands. The absence of a fixed schema and the utilization of decentralized architectures facilitate uninterrupted scalability (Sahatqija, et al., 2018).
5. **Unstructured or Semi-Structured Data:** In cases where the data comprises partially or completely structured formats, including content created by users, social networking posts, or data from sensors, NoSQL databases may provide a more suitable solution. The system facilitates storing and retrieving heterogeneous data categories without necessitating intricate schema alterations.

It is imperative to acknowledge that the selection of NoSQL across RDBMS necessitates a meticulous assessment of the particular prerequisites and limitations of the project. It is advisable to consider various factors, such as the necessity for data consistency, transactional prerequisites, and the existence of specific characteristics or tools of every kind for the database. Polyglot tenacity, which utilizes SQL and non-SQL databases, can be optimal in certain scenarios. This strategy capitalizes on each system's unique capabilities to cater to a program's distinct aspects (Lourenço, et al., 2015).

Tip

When choosing between NoSQL and RDBMS, consider factors like scalability, .performance, and data modeling requirements

To conclude, NoSQL databases have helped significantly transform information management by offering adaptable, expandable, and efficient options instead of conventional RDBMS. The rise and growth of NoSQL databases were primarily motivated by the necessity to manage vast amounts of data, ensure robust availability, and facilitate flexible development methodologies. NoSQL databases possess distinctive attributes and benefits, such as their adaptable data models, capacity for horizontal expansion, and ability to accommodate unstructured information, making them highly appealing for particular utilization scenarios. It is imperative to evaluate the prerequisites and limitations of the project before selecting NoSQL along with RDBMS, given that each database category possesses unique advantages and drawbacks. One can arrive at informed judgments under their data administration requirements by comprehending the incentives, attributes, and suitable application scenarios for NoSQL databases (Figure 2.8).

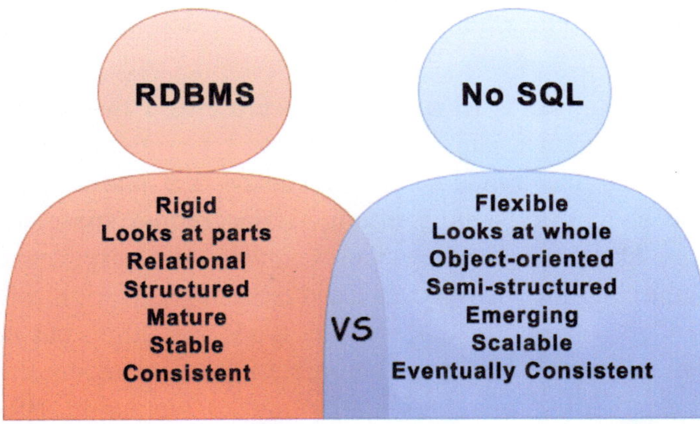

Figure 2.8. Illustration of difference NoSQL and RDBMS.

Source: Tatva soft, Creative Commons License.

2.6.4. Types of NoSQL Databases

2.6.4.1. Document-Oriented Databases

In contemporary times, documents-oriented databases have emerged as a prominent category of NoSQL databases, garnering substantial attention. These databases are engineered to efficiently store and handle data in a versatile and non-structured approach, rendering them appropriate for use in applications that experience dynamic data demands. Document-oriented records differ from relational databases in that they store information in self-describing papers, commonly represented in formats such as JSON or BSON, instead of organizing information with predetermined schemas (Lourenço, et al., 2015; Mason, 2015; Kumar & Mohanavalli, 2017; Vera, et al., 2015).

2.6.4.2. JSON and BSON Data Formats

JSON, which stands for JavaScript Object Notation, is an information exchange format that is both lightweight and easily readable by humans. The representation of structured information, such as keys and values, arrays, along with nested objects, is facilitated by a straightforward and user-friendly approach. JSON documents are frequently utilized in document-oriented databases, their ease of comprehension and broad support across various programming languages or systems. JSON's versatility and ease of use make it a highly suitable option for web-based applications requiring common data exchange among the user's device and the server (Nayak, et al., 2013).

BSON, also known as Binary JSON, represents a variant of the JSON format that incorporates supplementary data types and binary encoding. BSON preserves the fundamental characteristics and organization of JSON while concurrently augmenting its operational speed and storage effectiveness. The technology provides a condensed binary format for data, thereby minimizing the memory bandwidth on the network needed. BSON offers expanded support for diverse data types, including but not limited to dates. These data are binary and numeric types, enhancing the scope of expressive capabilities for representation and manipulation. BSON is the fundamental data structure in certain document-centric databases, such as MongoDB (Zhang, et al., 2014; Raut, 2017; Kaur & Rani, 2013).

2.6.4.3. MongoDB: Features and Data Modeling

MongoDB's free document-oriented file has gained significant popularity and widespread adoption. The software provides many functionalities, rendering it a flexible option for diverse applications. This study aims to examine MongoDB's fundamental characteristics and principles in conjunction with its methodology for data modeling.

1. **Flexible Schema:** The schema-less architecture of MongoDB enables adaptable and versatile data modeling. In contrast to relational databases that impose a strict schema framework, MongoDB permits documents to exhibit diverse structures within a single collection. The malleability of the data model allows developers to modify it by the evolution of the application, needing to necessitate intricate and protracted schema migrations (Reddy, et al., 2022).

2. **High Scalability:** The MongoDB database management system is specifically engineered to manage and accommodate extensive scalability demands effectively. Sharding splitting horizontal scaling by providing data across numerous computers or clusters. Through the process of sharding information, MongoDB is capable of effectively managing substantial datasets and accommodating intensive reading and writing workloads. Splitting can be set up using diverse criteria, including range-based splitting or hashed division, which enables effective data shipping and enhanced performance (Gupta, et al., 2017).
3. **Indexing and Querying:** The MongoDB platform offers comprehensive indexing functionalities designed to enhance query processing efficiency. The system supports various index types, encompassing single-field, substance, geospatial, and writing indexes designed to facilitate full-text search functionality. Indexes may affect the effectiveness of the database's search for pertinent documents, diminishing the necessity of scanning extensive collections. The optimization of indexes can have a substantial impact on the efficiency of data recovery processes.
4. **Replication and High Availability:** MongoDB provides inherent replication functionalities that guarantee elevated availability and data persistence. The replication process entails preserving numerous data duplicates across distinct servers, thereby establishing a set of replicas. In case of a malfunctioning elementary server, the backup set can autonomously select another main server to sustain reading and writing requests. The replication process facilitates the creation of duplicate copies of data, thereby ensuring redundancy and enabling the possibility of data recuperation in the event of hardware breakdowns or other catastrophic events (Tang & Fan, 2016).
5. **Data Modeling:** The process of data demonstration in MongoDB entails the creation of a well-designed framework for organizing and interconnecting documents to facilitate optimized query processing. Although document-driven databases such as MongoDB provide versatility in expressing data, it is crucial to devise the schema to enhance efficiency meticulously. A crucial factor is the denormalization of data pertains to the incorporation of interconnected data inside a solitary document or the replication of data across numerous documents. The implementation of

denormalization has the potential to enhance the efficiency of queries by mitigating the necessity for intricate join operations. However, it also presents difficulties in preserving data integrity throughout denormalized information.

The data modeling process within MongoDB is generally influenced by the particular methods of access along with query prerequisites of the application. The process entails the identification of crucial entities and their relationships, the determination of the structure of the document, and the optimization of indexes to enhance the efficiency of query execution. The malleability of MongoDB's structure affords the capacity for incremental and nimble data modeling, empowering programmers to hone and modify the data structure in tandem with the application's progression.

MongoDB's extensive range of features, such as its adaptable schema layout, capacity for high scaling, indexing functionalities, replication, and effective data modeling, render it a highly attractive option for uses that necessitate flexibility, scalability, and optimal performance. Comprehending the characteristics and optimal methodologies for modeling data within MongoDB is imperative to effectively exploit its potential in practical situations.

2.7. DATA MODELING IN NOSQL DATABASES

Data demonstration is of utmost importance in the design of files, as it establishes the structure and arrangement of data storage and retrieval within a given system. Although the basic concepts of modeling data to traditional RDBMS and NoSQL databases exist, notable distinctions exist in their methodologies (Oussous, et al., 2013). This part will provide an in-depth analysis of data demonstrated in NoSQL databases, highlighting the subtle distinctions between them and RDBMS.

2.7.1. Data Modeling Differences between NoSQL and RDBMS

The process of data demonstrating in RDBMS is centered on the relational framework, which entails organizing information into tables with predetermined schemas. The objective is to attain a database with a standard schema, mitigating duplication and guaranteeing data consistency. The establishment of relationships between tables is facilitated through the utilization of primary and foreign keys, which in turn allows for the efficient execution of joins when querying data. Conversely, NoSQL databases adopt an alternative framework to increase adaptability and scaling. In contrast to

RDBMS, NoSQL database systems are free from inflexible schemas, thereby facilitating the creation of schema-less or adaptable schema structures. The attribute of flexibility holds particular benefits in cases where the information structure is exposed to dynamic alterations or when managing partially or completely structured information (Vera-Olivera, et al., 2021).

2.7.2. Aggregation and Denormalization in NoSQL Databases

Databases using NoSQL exhibit a paradigm shift in emphasis from normalizing to denormalization accumulation. Normalization is a fundamental principle in RDBMS aimed at reducing redundancy. However, NoSQL databases tend to promote data denormalization to enhance the speed of queries.

The aggregation process entails consolidating interconnected data into a unified document or entity, expediting retrieval, and mitigating the necessity for intricate joins. In document-oriented databases such as MongoDB, the denormalization of related data inside a document obviates the need for expensive joins across numerous tables. Incorporating associated information immediately into a given document enables retrieval procedures to be executed in a solitary read, thereby augmenting efficiency and streamlining data accessibility (Han, et al., 2011).

Denormalization is a technique that facilitates the duplication of associated data across numerous documents and collections, thereby allowing for redundant data. Using redundant facilitates expedited queries and circumvents expensive joins, albeit at the cost of augmented storage necessities. The determination of whether or not to denormalize information stored in databases using NoSQL is contingent upon the particular query patterns and the given application's performance criteria (Atzeni, et al., 2020; Santos & Costa, 2016).

2.7.3. Schema-Less Design and Flexible Schemas

Non-relational databases, commonly called NoSQL databases, are characterized by their schema-less architecture, which sets them apart from traditional RDBMS. RDBMS have rigid schema enforcement, whereas non-relational databases (NoSQL) offer greater adaptability to data schema (Bhogal & Choksi, 2015).

NoSQL databases, including document-oriented files, exhibit heterogeneity in data structures inside a given collection or document. The adaptability of NoSQL databases to changing information needs is attributed to their adaptability, which obviates the necessity of modifying the database structure when fresh fields or characteristics are added. The flexibility of not being bound by predetermined schemas enables developers to incorporate or alter fields, thereby enhancing the agility of the creation process. In addition, the schema-less architecture of NoSQL databases provides a more efficient means of accommodating semi-structured and unstructured information (Zaki, 2014; Kaur & Rani, 2013). No-SQL database systems are capable of accommodating a wide range of data kinds and dynamic data models, rendering them highly suitable for functions that involve intricate or rapidly shifting data, owing to the absence of a predetermined structure that must be followed.

2.7.4. Consistency Models in NoSQL Databases

In the context of NoSQL databases, uniformity models specify how data is disseminated and harmonized across disparate systems. In contrast to RDBMS, which generally operate using robust uniformity as the failure mode, NoSQL database systems provide a variety of uniformity models to tackle the difficulties posed by distributed systems (Diogo, et al., 2019).

Certain NoSQL databases, including those utilizing key-value storage, adopt a consistent paradigm. In the present model, modifications performed on the database's contents can be promptly observed on the current replica. However, there may be a delay in disseminating these changes across other replicas. In due course, all duplicates attain a uniform state. The concept of eventual consistency confers the advantage of ensuring high reliability and tolerance for partitions, albeit at the cost of potential, albeit transient, data inconsistencies. Certain NoSQL databases offer robust consistency assurances that are comparable to those of RDBMS. The model guarantees strict data consistency by enabling immediate visibility of updates across all replicas. Attaining robust reliability in distributed computing environments frequently results in increased latency and decreased availability due to the requirement for coordination along with synchronization (Swaminathan, et al., 2016).

In addition to providing ultimately strong and weak consistency, several NoSQL databases present uniformity models on a spectrum among these two polarities. Causal uniformity and session alignment are instances of models that exemplify this notion. Causal consistency is a fundamental property that guarantees the preservation of causal dependencies by enforcing a strict ordering of causally related operations. Meeting consistency pertains to maintaining data consistency within a particular session or a client's perspective toward the data. This approach enables the provision of more robust assurances while simultaneously loosening the consistency across different sessions. The selection of a suitable reliability model is contingent upon the particular demands of an application. Determining the most appropriate model is significantly influenced by various factors, including but not limited to information access trends, scalability requirements, and the degree of tolerance for future consistency (Vera-Olivera, et al., 2021; Banerjee & Sarkar, 2016).

2.8. SCALABILITY AND PERFORMANCE CONSIDERATIONS

2.8.1. Scalability Challenges in NoSQL Databases

The scalability issue holds significant importance in contemporary information management systems, particularly when considering applications of extensive magnitude and data of substantial size. NoSQL databases are engineered to manage extensive volumes of data and offer horizontal scalability, enabling further servers or nodes to spread the data along with the workload. Notwithstanding, attaining scalability within NoSQL databases presents distinct challenges (Pokorny, 2011).

Remember

Horizontal scaling and sharding are important techniques used to address scalability .in NoSQL databases

Data distribution poses a significant scalability problem for NoSQL databases. In contrast to conventional relational database systems that depend on vertical expansion to enhance the computing power of one server, NoSQL databases can employ horizontal expansion to disseminate data across numerous servers. The present distribution of data presents challenges in maintaining consistent information and optimizing data access efficiency.

When data is dispersed among many servers, it's imperative to consider the partitioning of information and queries through the relevant nodes. Maintaining uniform distribution of data across fragments or divisions is imperative to avoid the occurrence of hotspots, wherein specific nodes are burdened with an inequitable share of data along with queries. Attaining equitable data distribution necessitates meticulous deliberation of the fragment key, which ascertains how information is partitioned (Mapanga & Kadebu, 2013).

Consistency in data poses an additional difficulty. NoSQL databases frequently prioritize the attributes of high reliability and tolerance for partitions, as per the capacity availability theorem, over strict consistency. The compromise above facilitates enhanced operational efficiency and resilience within distributed systems, albeit at the expense of eventual consistency, which may result in momentary discrepancies among data replicas. Achieving optimal performance over an application while maintaining consistency requirements necessitates finding an appropriate equilibrium (Sareen & Kumar, 2015; Khan, et al., 2023).

2.8.2. Horizontal Scaling and Sharding

Implementing horizontal growth is a fundamental strategy for scaling databases using NoSQL. The process entails the dispersion of data among numerous servers or just nodes to manage augmented workloads and quantities. Sharding is a commonly employed method in which a given dataset is partitioned into fewer subsets, called shards or divisions, and subsequently stored on distinct servers.

The partitioning process entails the selection of a shard key that governs the partitioning of data. The selection of the shard key necessitates meticulous consideration to ensure the equitable dispersion of data among shards and circumvent the occurrence of hotspots or disproportionate workloads. Typical strategies employed for fragmented key selection encompass utilizing a unique identification or a field that denotes the locality of the data, such as geographical location. Implementing splitting yields various advantages, such as the heightened capacity for storage, enhanced reading and writing performance, and faults. Through data distribution, individual shards can manage a specific subset of the overall workload, facilitating concurrent processing and mitigating contention. Furthermore, in the event of a shard along with server failure, the remaining fragments can persistently provide data, guaranteeing high availability (Patil, et al., 2017; Dhulavvagol & Totad, 2023).

Nevertheless, the implementation of sharding also presents certain difficulties. Co-located data, which refers to data that must be accessed in conjunction, may be distributed across multiple shards, necessitating further synchronization and potentially compromising system efficiency. The effective scaling of a system necessitates meticulous observation and realignment of shards to sustain a uniform dispersion of information and workload (Hashim, et al., 2022; Abdelhafiz & Elhadef, 2021).

2.8.3. CAP theorem and Trade-offs in Distributed Systems

Brewer's theorem, commonly called the theory of CAP, posits that a networked data system cannot feasibly offer reliability, accessibility, and partition tolerance concurrently. Consistency refers to the property of all nodes perceiving identical information, while availability denotes the capacity of the infrastructure to respond to all requests, even in the face of errors. Partition tolerance, upon the other hand, pertains to the system's capacity to continue functioning despite partitions within the network (Ahmad, et al., 2020; Diack, et al., 2013; Garg, 2022) (Figure 2.9).

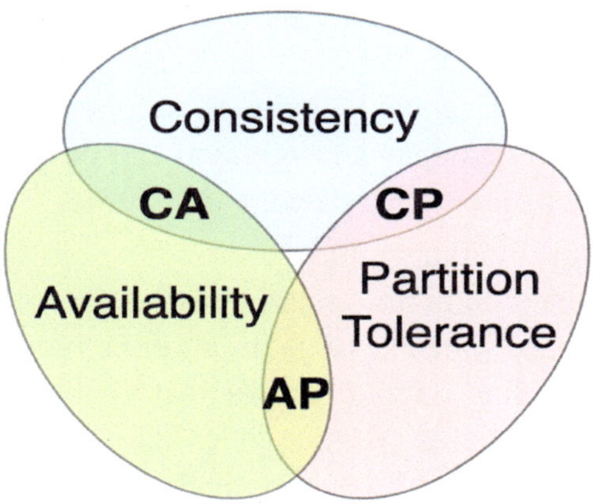

Figure 2.9. Schematic of CAP theorem.

Source: JamieMcCarthy, Creative Commons License.

NoSQL databases are primarily intended for scaling and high availability and therefore tend to put tolerance for partitions and availability across strict consistency. In the event of network divisions or mistakes, continued accessibility and tolerance for partitions over consistency. The trade-off above gives rise to distinct consistency models, including eventual or related consistency, wherein data replicas may undergo temporary divergence but ultimately converge (Gilbert & Lynch, 2012).

Comprehending the trade-offs linked with the concept of CAP is crucial in designing and utilizing NoSQL databases. The selection of consistency models may differ based on the particular demands of an application. Certain programs require eventual uniformity, whereas others necessitate strong consistency assurances. Selecting a suitable consistency framework for a given use case requires architects and developers to evaluate the associated compromises thoroughly.

2.8.4. Performance Tuning in NoSQL Databases: Caching, Indexing, and Partitioning

A range of methodologies can be implemented to enhance the efficiency of databases that use NoSQL, such as caching, searching, and partitioning.

- **Caching**: The caching process entails retaining frequently used data near the program in memory, aiming to mitigate latency and enhance response times. Incorporating cached information in memory systems, like Redis or Memcached, alongside NoSQL relational databases can facilitate expedited retrieval of frequently utilized data. Caching techniques like the most recently executed (LRU) or time-dependent removal can be utilized to administer the cache proficiently (Hsieh, et al., 2011).
- **Indexing:** Indexing constitutes a pivotal facet of optimizing performance. Similar to relational databases, databases built with NoSQL employ indexes to facilitate effective information retrieval by constructing data structures that enable rapid lookup based on particular fields. Various indexing methods can be employed based on the type of database, including B-trees and LSM-trees. It is advisable to exercise caution while choosing suitable fields for indexed strike a balance between query execution efficiency and the upkeep overhead of the index (Pandey, et al., 2018).
- **Partitioning:** As previously mentioned, partitioning refers to fragmenting the dataset into smaller parts or partitions. Effective partitioning techniques can enhance system efficiency by providing information and workload among numerous servers. Selecting a suitable shard key and preventing hotspots or uneven partitions are critical factors in achieving effective data sharing and reducing bottlenecks.

Apart from caches, indexing, and splitting, additional techniques for enhancing performance encompass optimization of queries, data denormalization, however and appropriate hardware provisioning. The query optimization process entails examining and enhancing question execution schedules from responses. The deliberate introduction of redundant data, known as data denormalization, can enhance read performance while incurring higher storage demands. The general efficiency of NoSQL tables can be significantly influenced by appropriate hardware provisioning, which may involve the utilization of solid-state disks (SSDs) or optimizing network arrangements (Chawla, et al., 2020).

By implementing performance optimization strategies, entities can guarantee that their MySQL databases can effectively manage extensive workloads and deliver the necessary level of responsiveness to satisfy the requirements of contemporary applications.

2.9. ADVANCED SQL TECHNIQUES

2.9.1. Advanced Querying

The utilization of advanced searching methods in SQL empowers users to extract and handle data more intricately than conventional SELECT declarations. This section delves into three crucial facets of sophisticated querying: subqueries, correlated queries, prevalent table phrases (CTEs), and recurrent queries, which are instrumental in managing hierarchical data. Subqueries are used, and associated queries are potent instruments that enable the nesting of queries inside additional queries. A subquery refers to a query incorporated within the primary query and may be utilized in different sections of the SQL query, including the SELECT clause and the WHERE section. Subqueries offer a means to execute computations, filtering, or collection on a certain amount of data before the execution of the primary query. An illustrative scenario involves retrieving employees whose salaries exceed the mean salary of their respective departments. One possible approach to accomplish this task is to employ a subquery that computes the average wage for each division and subsequently contrasts this with the median wage of every worker in the primary query (Woods, et al., 2014).

Correlated queries extend the functionality of subqueries by incorporating references to columns derived from the exterior query inside the subquery. The correlation above facilitates the subquery's dependence on the outcomes of the exterior query. Correlated searches are valuable for conducting row-by-row contrasts or filtering according to principles from the outside query. Subsequently, we shall examine the ubiquitous table expressions known as Common Table Expressions (CTEs). CTEs are a type of short-term lead set that may be utilized within a query. They are particularly advantageous in situations where intricate and repeatable subqueries are necessary. CTEs enhance the comprehensibility and sustainability of code by enabling the definition of a designated temporary following table expression which can be invoked repeatedly in a query.

The initial step in utilizing a CTE involves its definition through the WITH clause, wherein the CTE's name and column definitions are specified. The CTE is subsequently alluded to within the primary query in a manner that resembles a conventional table or view. CTEs have the potential to enhance the efficiency of intricate queries by decomposing them into more comprehensible segments, thereby facilitating the reuse of code. An instance of utilizing a CTE is when there is necessary to identify the highest-selling commodities within every category. One may generate a CTE that employs sales volume to rank products within respective categories and subsequently merge the CTE alongside the primary query to obtain the intended outcomes (Olsen, 2000).

Recursive queries represent an advanced SQL methodology employed to query hierarchical data. Hierarchical information depicts connections among entities according to a hierarchical arrangement, such as those found in genealogy databases or organizational diagrams. Recursive queries facilitate the traversal of a hierarchy by enabling the navigation of parent-child connections data. To execute a recursive inquiry, it is customary to employ a CTE that refers to its structure. The recursive component of the inquiry encompasses the primary inquiry seed, also referred to as the anchor member, and the recursive expression that specifies how the inquiry should advance by merging with the CTE itself. As mentioned, the procedure persists until a specified termination criterion is satisfied.

As an illustration, consider a tabular representation of a corporate structure featuring attributes such as employee identification number, name, and manager identification number. Formulating a recursive query to retrieve comprehensive information on employees and their interpersonal connections is possible. This query would commence at the highest level of the organizational hierarchy, namely the CEO, and proceed downwards through the lowest-level staff members (Rivero & Ruiz, 2002).

To conclude, utilizing advanced SQL searching techniques offers potent means for modifying and finding data more intricately. Subqueries and associated queries facilitate the nesting of queries, thereby enabling intricate computations and filtration. Using CTEs can improve the comprehensibility and adaptability of queries by generating provisional outcome sets. Recursive queries are highly useful in the context of searching structures of data that exhibit a hierarchical organization. Proficiency in advanced SQL techniques can enhance one's query abilities while allowing the efficient handling of intricate retrieval of information tasks.

2.9.1.1. Advanced Indexing

The indexing process is paramount in enhancing the efficiency and effectiveness of database management systems. Creating an organized data version facilitates efficient data extraction, enabling expedited access according to particular search criteria. The upcoming chapter will delve into diverse advanced indexed methodologies that have the potential to augment the efficiency of databases substantially. The present study will explore various indexing methods that can be employed to enhance system performance. Specifically, the study will examine the benefits of bitmap indices, full-text indexed along with search, and geographic indexed as applied to data related to geography (Li, et al., 2005).

2.9.1.2. Indexing Strategies for Improved Performance

Indexing is a fundamental methodology employed in databases to enhance the expediency of data retrieval. B-trees and other conventional indexing methods are extensively employed and offer effective data retrieval in various contexts. Nevertheless, supplementary indexing techniques can enhance performance even further in the case of extensive databases that entail complex queries and information types (Habimana, 2015).

A strategy that can be employed is the utilization of combined indexes, which entail the establishment of an index on several columns as opposed to a solitary column. The relational database can effectively leverage the combined index to fulfill queries with multiple search criteria through this approach. Implementing this technique can significantly reduce the number of data accesses needed, yielding expedited execution of query durations (Zhou, et al., 2012).

An additional approach to indexing is the implementation of function-based indexes. The methodology generates an index by executing an operator or conveying on the dataset. Using function-based indexes can prove advantageous, especially in scenarios involving intricate data manipulations or computations. Through precomputation and indexing, the relational database efficiently retrieves the needed information without incurring the expenses associated with on-the-fly calculation (Seshadri & Paskin, 1997).

2.9.1.3. Bitmap Indexes and Their Advantages

Bitmap indexing is a distinct methodology that can offer considerable performance advantages in specific situations. In contrast to conventional indices that retain value suggestions to the associated data, bitmap indices maintain a bitmap for every unique value in a given column. The bitmap utilizes individual bits to indicate the presence or absence of a particular value within a given row.

Bitmap indexes offer a notable benefit in effectively managing queries encompassing various criteria. The bitwise operations applied to the bitmaps are the relational database expeditiously ascertains the rows that meet the specified search parameters. The implementation of this technique has the potential to significantly enhance the speed of query execution, particularly in cases where intricate queries encompassing numerous predicates are involved (Wu, et al., 2001).

Bitmap indices are highly appropriate for datasets that exhibit low cardinality, wherein the total amount of unique values in a given column is relatively diminutive compared with the overall number of rows. In instances of this nature, bitmap indexes possess the potential for significant compression, thereby reducing storage demands. Furthermore, bitmap indexes exhibit high efficiency in managing set operations, including union, connection, and distinction, rendering them appropriate for analytical inquiries and data warehousing contexts (Wu, et al., 2008).

2.9.1.4. Full-Text Indexing and Search

Conventional indexing methods are predominantly centered on precise matches, rendering them less efficacious for textual data that necessitates keyword searches and using natural languages in query formulation. The limitation of inefficient searching inside text fields can be overcome by utilizing full-text indexed search methods, which enable the efficient retrieval of information using phrases or words.

Full-text indexing entails generating a database of textual information, such as items or documents, by examining the content and identifying pertinent terms. This index facilitates expedient searches based on keywords, thereby enabling prompt retrieval of relevant documents in response to user queries. To accomplish this task, full-text indexing employs stemming, tokenization, and stop-word elimination techniques to produce a condensed index framework (Vyverman, et al., 2012; Melnik, et al., 2001).

Similar text analysis methods are employed to process the query when conducting a full-text search. The process involves matching the query terms with the indexed terms, following which the search engine generates a list of the most pertinent documents. The order of importance of these documents is determined by various factors such as frequency range, proximity, and significance ranking computation.

The utilization of full-text indexed and search methodologies is prevalent in diverse applications, such as archives of documents, content management systems, and search engines. Implementing these methods results in improved user experience and enhanced performance of applications that handle textual information, facilitating efficient written investigating (Ferragina, et al., 2007).

2.9.1.5. Spatial Indexing and Geographic Data

Spatial searching is a distinct indexing methodology employed to preserve and access geographical information effectively. Using spatial data management systems facilitates the efficient processing of inquiries about spatial entities such as points, lines, and polygons. This capability empowers software applications to execute functions such as location-based searching, closeness analysis, and graphical representation of geographic information (Jones, et al., 2004).

This frequently employed spatial indexing method utilizes a tree-based data structure to facilitate data indexing in multiple dimensions. The R-tree is a hierarchical data structure that arranges spatial entities in a tree-like form. Each node within the tree denotes a leaping geographical area, a group of entities. Utilizing a hierarchical structure facilitates expedient spatial search procedures, diminishing the number of entities necessitating scrutiny throughout processing queries.

Spatial indexing facilitates spatial inquiries such as range queries, closest neighbor inquiries, and spatial unions. The software offers efficient functionality for managing intricate spatial operations such as intersection, confinement, and separation computations. The operations above hold significant importance in applications of geographic information systems (GIS), navigational systems, and services that use location (Zhu, et al., 2007).

The significance of geographic information has risen across multiple fields, such as travel, urban planning, and ecological sciences. Consequently, spatial indexed has become an essential component of contemporary database systems. The tool's capacity to enhance the analysis of geographic queries renders it an invaluable tool for proficiently handling and evaluating geographic information.

To summarize, utilizing advanced indexed techniques is crucial in augmenting the efficiency of databases. Utilizing techniques that include composite indexes and function-based indices can enhance the efficiency of executing queries and facilitate intricate data transformations. Bitmap indexes offer benefits in efficiently managing various circumstances and set activities. Full-text indexing facilitates expedited searches based on keywords within textual data, whereas spatial indexing streamlines the handling of geographical data queries. Using sophisticated indexing methodologies, databases can enhance the speed and efficacy of data retrieval, ultimately improving the system's effectiveness (Fang, et al., 2008).

2.9.2. Transactions and Concurrency Control

The proper management of transactions, along with control over concurrency, is of utmost importance in upholding data integrity and preserving an information system's coherence. This section will delve into the multifaceted components of deals and cooperation control. Specifically, it will examine the ACID characteristics of transactions, cooperation control techniques, including securing and MVCC, separation levels, and approaches for averting and handling deadlocks (Barghouti, 1991).

2.9.2.1. ACID Properties of Transactions

ACID is a condensed form of Atomicity, Consistency, and Isolation. The properties above establish the fundamental attributes of an agreement and guarantee its dependability and accuracy.

The principle of atomicity ensures that an activity is regarded as an indivisible entity of operation. The statement implies that the transaction's modifications are either entirely stored in the database's structure or not. In the event of a transactional failure, the entirety of the transaction is reverted, resulting in the database retaining its original state (Abbas, et al., 2016).

Maintaining consistency guarantees that a given transaction results in the transition of the relational database from an initial in-line state to a final in-keeping state. The database is ensured to maintain its validity throughout the transaction by enforcing integrity limitations, such as uniqueness constraints or foreign key relationships. The concept of isolation pertains to the ability of concurrent transactions to execute simultaneously without interference. The implementation of isolation guarantees the absence of interference among transactions. It ensures that every transaction perceives a coherent database perspective, notwithstanding other transactions' concurrent modification of identical data. The attribute of durability ensures that the outcomes of a transaction remain intact and enduring even in the event of any subsequent system failures, thereby providing a guarantee of permanence. The alterations are retained in a persistent storage medium-sized, such as a hard drive, and can be retrieved during a system malfunction or power outage (Xie, et al., 2015).

2.9.2.2. Concurrency Control Mechanisms: Locking and MVCC

Concurrency controls are utilized to manage scenarios where multiple transactions simultaneously have access to and alter identical data. Securing along with Multi-Version Concurrency Control are two frequently employed mechanisms. The locking mechanism is extensively employed to ensure data consistency by providing sole access to a particular resource. Lock acquisition can occur at varying levels, including a database, table, and row level. For an operation to gain access to a particular resource, it is necessary to obtain the corresponding lock. If a different transaction already holds a lock, the requester transactions may be required to wait or stop until the security key is released. By keeping several copies of a piece of information, MVCC, on the opposite hand, enables simultaneous viewing and modification of data by numerous transactions. A picture of the database's structure as it was at the time of the operation is shown for each transaction. Before they are committed, additional transactions aren't impacted by changes made by one. With this method, concurrency is increased, and explicit barring is unnecessary (Di Sanzo, et al., 2008; Neumann, et al., 2015).

2.9.2.3. Isolation Levels and Their Impact on Performance

The concept of isolation levels pertains to the extent of isolation among tandem transactions and the scope of visibility that modifications performed by a single transaction have for other transactions. Various isolation levels are available, such as Read Uncommitted, Serializable, Repeatable Read, Read Committed, or presenting a distinct proportion of multitasking.

The Read Uncommitted separation level is the least restrictive isolation level, allowing a transaction to access and read indifferent modifications made by other concurrent transactions. Although it offers superior concurrency, this approach compromises data uniformity. The Read

Committed isolation level ensures that a transaction can solely access committed modifications carried out by other operations, thereby preventing the occurrence of dirty reads. Repeatable The Read operation guarantees that the identical query executed repetitively yields coherent outcomes in the context of a transaction, notwithstanding any alterations made to the data by subsequent transactions. The Serializable feature guarantees the utmost level of information uniformity by ensuring that operations happen in a manner that mimics serial execution. However, this feature may lead to reduced concurrency due to heightened locking (Larson, et al., 2011).

The selection of the separation level is contingent upon the particular demands of the application in question. Elevated levels of isolation guarantee the uniformity of data. However, they may hurt cooperation and efficiency. The judicious selection of the suitable isolation level is crucial in achieving an optimal trade-off between maintaining consistent data and enhancing system performance (Harding, et al., 2017).

2.9.2.4. Deadlocks and Deadlock Prevention Strategies

A state of deadlock arises in a system of concurrent transactions if multiple transactions are mutually blocked and indefinitely awaiting one another to relinquish resources. Deadlocks can cause a system to cease functioning and necessitate intervention for resolution. To avert deadlocks, diverse tactics can be implemented. A viable strategy involves the implementation of a resource-requesting protocol, wherein transactions are designed to solicit and relinquish resources in a pre-established sequence. As mentioned earlier, the action effectively eradicates the potential for circular waits, a prevalent instigator of deadlock occurrences. An additional approach involves implementing timeouts, whereby a transaction that fails to obtain an object within a designated period relinquishes its presently held materials and attempts to acquire them again later. Deadlock detection programs might be utilized to periodically examine the existence of deadlocks and implement suitable measures for resolving them.

Concurrency controls, such as locking along with MVCC, prevent impasses by regulating the acquisition of resources by deals. Deadlocks may be prevented or reduced by providing exclusive utilization of resources (Agrawal, et al., 1987).

Incorporating transactions and cooperation control mechanisms is imperative for establishing a resilient and dependable database infrastructure. Comprehending the ACID characteristics of transactions, cooperation management techniques such as locking along with MVCC, separation levels, and techniques for preventing deadlocks is imperative for devising and administering database systems that exhibit superior performance. Transactions and cooperation control mechanisms are crucial for efficiently operating databases in a multi-user and concurrent environment. These mechanisms ensure the accuracy of data and uniformity and avoid deadlocks (Tang, et al., 2012).

Activity 2.1

Imagine you are designing a new web application that requires storing user data, including preferences and activity logs. Based on the requirements of scalability and flexible schemas, would you choose a relational database (RDBMS) or a NoSQL .database? Justify your choice

2.10. SUMMARY

In this chapter on the fundamentals of data management, we explore the key concepts and principles that form the foundation of efficient data handling. The chapter begins by introducing relational databases, which are widely used in various applications. We delve into the definition of relational databases and highlight their essential characteristics.

One fascinating factual stat (Did you know?) about relational databases is that SQL, or SQL, serves as the standard language for managing them. SQL empowers users to interact with relational databases, allowing them to retrieve, insert, update, and delete data. Remember that SQL encompasses both DDL and DML statements, which provide the means to define and manipulate the structure and content of databases, respectively.

To reinforce this knowledge, we offer a helpful tip: practice writing basic SQL queries such as SELECT, INSERT, UPDATE, and DELETE. By honing your SQL skills, you can confidently navigate relational databases and perform essential data management tasks.

Moving on, we shift our attention to NoSQL databases, which have emerged as an alternative to traditional relational databases due to specific needs and motivations. One intriguing factual stat (Did you know?) about NoSQL databases is that their development was driven by the desire for more flexible and scalable data storage solutions. These databases offer unique characteristics and advantages that cater to diverse use cases.

To ensure a comprehensive understanding of NoSQL databases, we remind you to consider the fundamental differences in data modeling between NoSQL and RDBMS. Additionally, we provide a useful tip: when faced with the decision of choosing between NoSQL and RDBMS, carefully assess factors such as scalability, performance, and data modeling requirements. This evaluation will guide you toward selecting the most suitable database technology for your specific needs.

Furthermore, the chapter explores scalability and performance considerations in depth. We highlight the scalability challenges that arise in NoSQL databases due to their distributed nature. Remember that horizontal scaling and sharding are essential techniques used to address these challenges. To enhance the performance of NoSQL databases, we offer a valuable tip: implement strategies such as caching, indexing, and partitioning to optimize data access and retrieval. To conclude the chapter, we touch upon advanced SQL techniques, including advanced querying and transactions. However, it's important to note that the chapter summary doesn't include the specific chapter number. By comprehending the fundamentals of data management, relational databases, NoSQL databases, and SQL techniques, you'll be equipped with the knowledge and skills necessary to make informed decisions regarding data storage, retrieval, and manipulation.

REVIEW QUESTIONS

1. What is the purpose of SQL in relational databases?
2. Name two types of SQL statements.
3. Explain the importance of normalization in database design.
4. What are the different normalization forms?
5. How does denormalization impact database design?
6. What are the motivations behind the development of NoSQL databases?
7. List two advantages of using NoSQL databases.
8. What are the differences in data modeling between NoSQL and RDBMS?
9. How do scalability challenges differ in NoSQL databases compared to RDBMS?
10. Name three performance optimization techniques used in NoSQL databases.

REFERENCES

1. Abbas, Q., Shafiq, H., Ahmad, I., & Tharanidharan, S., (2016). Concurrency control in distributed database system. In: *2016 International Conference on Computer Communication and Informatics (ICCCI)* (pp. 1–4). IEEE.
2. Abdelhafiz, B. M., & Elhadef, M., (2021). Sharding database for fault tolerance and scalability of data. In: *2021 2nd International Conference on Computation, Automation and Knowledge Management (ICCAKM)* (pp. 17–24). IEEE.
3. Aboutorabi[a], S. H., Rezapour, M., Moradi, M., & Ghadiri, N., (2015). Performance evaluation of SQL and MongoDB databases for big e-commerce data. In: *2015 International Symposium on Computer Science and Software Engineering (CSSE)* (pp. 1–7). IEEE.
4. Abyzov, A. N., (2023). Balancing data normalization and denormalization in sports competition management platforms: A comparative analysis. *Инновации и Инвестиции*, (4), 273–277.
5. Agrawal, R., Carey, M. J., & McVoy, L. W., (1987). The performance of alternative strategies for dealing with deadlocks in database management systems. *IEEE Transactions on Software Engineering*, (12), 1348–1363.
6. Agrawal, S., Chaudhuri, S., Kollar, L., Marathe, A., Narasayya, V., & Syamala, M., (2005). Database tuning advisor for Microsoft SQL server 2005. In: *Proceedings of the 2005 ACM SIGMOD International Conference on Management of Data* (pp. 930–932).
7. Ahmad, T., Morelli, U., & Ranise, S., (2020). Deploying access control enforcement for IoT in the cloud-edge continuum with the help of the CAP theorem. In: *Proceedings of the 25th ACM Symposium on Access Control Models and Technologies* (pp. 213–220).
8. Albarak, M., & Bahsoon, R., (2018). Prioritizing technical debt in database normalization using portfolio theory and data quality metrics. In: *Proceedings of the 2018 International Conference on Technical Debt* (pp. 31–40).
9. Amornchewin, R., (2018). The development of SQL language skills in data definition and data manipulation languages using exercises with Quizizz for students' learning engagement. *Indonesian Journal of Informatics Education, 2*(2), 85–90.

10. Armbrust, M., Xin, R. S., Lian, C., Huai, Y., Liu, D., Bradley, J. K., & Zaharia, M., (2015). Spark SQL: Relational data processing in spark. In: *Proceedings of the 2015 ACM SIGMOD International Conference on Management of Data* (pp. 1383–1394).
11. Atzeni, P., Bugiotti, F., Cabibbo, L., & Torlone, R., (2020). Data modeling in the NoSQL world. *Computer Standards & Interfaces, 67*, 103149.
12. Bahmani, A. H., Naghibzadeh, M., & Bahmani, B., (2008). Automatic database normalization and primary key generation. In: *2008 Canadian Conference on Electrical and Computer Engineering* (pp. 000011–000016). IEEE.
13. Banerjee, S., & Sarkar, A., (2016). Modeling NoSQL databases: From conceptual to logical level design. In: *3rd International Conference Applications and Innovations in Mobile Computing (AIMoC 2016)* (pp. 10–12). Kolkata, India.
14. Barghouti, N. S., & Kaiser, G. E., (1991). Concurrency control in advanced database applications. *ACM Computing Surveys (CSUR), 23*(3), 269–317.
15. Beaubouef, T., Petry, F. E., & Ladner, R., (2005). Normalization in a rough relational database. In: *Rough Sets, Fuzzy Sets, Data Mining, and Granular Computing: 10th International Conference, RSFDGrC 2005, Regina, Canada, August 31-September 3, 2005, Proceedings, Part I 10* (pp. 275–282). Springer Berlin Heidelberg.
16. Berry, M. A., & Linoff, G. S., (2000). Mastering data mining: The art and science of customer relationship management. *Industrial Management & Data Systems, 100*(5), 245–246
17. Bhogal, J., & Choksi, I., (2015). Handling big data using NoSQL. In: *2015 IEEE 29th International Conference on Advanced Information Networking and Applications Workshops* (pp. 393–398). IEEE.
18. Bock, D. B., (1997). Entity-relationship modeling and normalization errors. *Journal of Database Management (JDM), 8*(1), 4–13.
19. Bondiombouy, C., & Valduriez, P., (2016). Query processing in multistore systems: An overview. *International Journal of Cloud Computing, 5*(4), 309–346.
20. Buckles, B. P., & Petry, F. E., (1982). A fuzzy representation of data for relational databases. *Fuzzy Sets and Systems, 7*(3), 213–226.

21. Ceri, S., & Gottlob, G., (1986). Normalization of relations and prolog. *Communications of the ACM, 29*(6), 524–544.
22. Chaudhuri, S., & Narasayya, V. R., (1997). An efficient, cost-driven index selection tool for Microsoft SQL server. In: *VLDB* (Vol. 97, pp. 146–155).
23. Chawla, T., Singh, G., Pilli, E. S., & Govil, M. C., (2020). Storage, partitioning, indexing and retrieval in big RDF frameworks: A survey. *Computer Science Review, 38*, 100309.
24. Chen, T. X., Liu, S. S., Meyer, M. D., & Gotterbarn, D., (2007). An introduction to functional independency in relational database normalization. In: *Proceedings of the 45th Annual Southeast Regional Conference* (pp. 221–225).
25. Colley, D., & Stanier, C., (2017). Identifying new directions in database performance tuning. *Procedia Computer Science, 121,* 260–265.
26. Date, C. J., (1988). Defining data types in a database language. *ACM SIGMOD Record, 17*(2), 53–76.
27. Demba, M., (2013). Algorithm for relational database normalization up to 3NF. *International Journal of Database Management Systems, 5*(3), 39.
28. Dhabe, P. S., Patwardhan, M. S., Deshpande, A. A., Dhore, M. L., Barbadekar, B. V., & Abhyankar, H. K., (2010). Articulated entity relationship (AER) diagram for complete automation of relational database normalization. *International Journal of Database Management Systems, 2*(2), 84–100.
29. Dhulavvagol, P. M., & Totad, S. G., (2023). Performance enhancement of distributed system using HDFS federation and sharding. *Procedia Computer Science, 218*, 2830–2841.
30. Di Sanzo, P., Ciciani, B., Quaglia, F., & Romano, P., (2008). A performance model of multi-version concurrency control. In: *2008 IEEE International Symposium on Modeling, Analysis and Simulation of Computers and Telecommunication Systems* (pp. 1–10). IEEE.
31. Diack, B. W., Ndiaye, S., & Slimani, Y., (2013). CAP theorem between claims and misunderstandings: What is to be sacrificed. *International Journal of Advanced Science and Technology, 56*, 1–12.
32. Diogo, M., Cabral, B., & Bernardino, J., (2019). Consistency models of NoSQL databases. *Future Internet, 11*(2), 43.

33. Dongare, Y. V., Dhabe, P. S., & Deshmukh, S. V., (2011). *RDBNorma:-A Semi-Automated Tool for Relational Database Schema Normalization Up to Third Normal Form*. arXiv preprint arXiv:1103.0633.
34. Eisenberg, A., & Melton, J., (2002). SQL/XML is making good progress. *ACM SIGMOD Record, 31*(2), 101–108.
35. Embley, D. W., & Ling, T. W., (1989). *Synergistic Database Design with an Extended Entity-Relationship Model* (pp. 111–128). In ER.
36. Fang, Y., Friedman, M., Nair, G., Rys, M., & Schmid, A. E., (2008). Spatial indexing in Microsoft SQL server 2008. In: *Proceedings of the 2008 ACM SIGMOD International Conference on Management of Data* (pp. 1207–1216).
37. Ferragina, P., Manzini, G., Mäkinen, V., & Navarro, G., (2007). Compressed representations of sequences and full-text indexes. *ACM Transactions on Algorithms (TALG), 3*(2), 20-es.
38. Garg, D., (2022). *Cost and Performance Trade-off Evaluation in Microservices Impacted by the CAP Theorem Limitations*. Doctoral dissertation.
39. Gilbert, S., & Lynch, N., (2012). Perspectives on the CAP theorem. *Computer, 45*(2), 30–36.
40. Gupta, A., Tyagi, S., Panwar, N., Sachdeva, S., & Saxena, U., (2017). NoSQL databases: Critical analysis and comparison. In: *2017 International Conference on Computing and Communication Technologies for Smart Nation (IC3TSN)* (pp. 293–299). IEEE.
41. Habimana, J., (2015). Query optimization techniques-tips for writing efficient and faster SQL queries. *International Journal of Scientific & Technology Research, 4*(10), 22–26.
42. Han, J., Haihong, E., Le, G., & Du, J., (2011). Survey on NoSQL database. In: *2011 6th International Conference on Pervasive Computing and Applications* (pp. 363–366). IEEE.
43. Harding, R., Van, A. D., Pavlo, A., & Stonebraker, M., (2017). An evaluation of distributed concurrency control. *Proceedings of the VLDB Endowment, 10*(5), 553–564.
44. Hashim, F., Shuaib, K., & Zaki, N., (2022). Sharding for scalable blockchain networks. *SN Computer Science, 4*(1), 2.
45. Helland, P., (2019). The trade-offs between write and read. *Communications of the ACM, 62*(11), 111–113.

46. Hillenbrand, A., Störl, U., Levchenko, M., Nabiyev, S., & Klettke, M., (2020). Towards self-adapting data migration in the context of schema evolution in NoSQL databases. In: *2020 IEEE 36th International Conference on Data Engineering Workshops (ICDEW)* (pp. 133–138). IEEE.
47. Hsieh, M. J., Chang, C. R., Ho, L. Y., Wu, J. J., & Liu, P., (2011). SQLMR: A scalable database management system for cloud computing. In: *2011 International Conference on Parallel Processing* (pp. 315–324). IEEE.
48. Hussain, T., Shamail, S., & Awais, M. M., (2003). Eliminating process of normalization in relational database design. In: *7th International Multi Topic Conference, 2003: INMIC 2003* (pp. 408–413). IEEE.
49. Imieliński, T., & Lipski, Jr. W., (1984). Incomplete information in relational databases. *Journal of the ACM (JACM), 31*(4), 761–791.
50. Jahan, A., & Edwards, K. L., (2015). A state-of-the-art survey on the influence of normalization techniques in ranking: Improving the materials selection process in engineering design. *Materials & Design (1980–2015), 65*, 335–342.
51. Jones, C. B., Abdelmoty, A. I., Finch, D., Fu, G., & Vaid, S., (2004). The SPIRIT spatial search engine: Architecture, ontologies, and spatial indexing. In: *Geographic Information Science: Third International Conference, GIScience 2004, Adelphi, MD, USA, October 20–23, 2004; Proceedings 3* (pp. 125–139). Springer Berlin Heidelberg.
52. Kamatkar, S. J., Kamble, A., Viloria, A., Hernández-Fernandez, L., & Cali, E. G., (2018). Database performance tuning and query optimization. In: *Data Mining and Big Data: Third International Conference, DMBD 2018, Shanghai, China, June 17–22, 2018, Proceedings 3* (pp. 3–11). Springer International Publishing.
53. Kanade, A., Gopal, A., & Kanade, S., (2014). A study of normalization and embedding in MongoDB. In: *2014 IEEE International Advance Computing Conference (IACC)* (pp. 416–421). IEEE.
54. Kau, S. C., & Tseng, J. C. R., (1994). MQL-a query language for multimedia database. In: *5th IEEE COMSOC International Workshop on Multimedia Communications* (pp. 5–1). IEEE.
55. Kaur, K., & Rani, R., (2013). Modeling and querying data in NoSQL databases. In: *2013 IEEE International Conference on Big Data* (pp. 1–7). IEEE.

56. Kearns, R., Shead, S., & Fekete, A., (1997). A teaching system for SQL. In: *Proceedings of the 2nd Australasian Conference on Computer Science Education* (pp. 224–231).
57. Kemper, A., Kossmann, D., & Zeller, B., (1999). Performance tuning for SAP R/3. *IEEE Data Eng. Bull., 22*(2), 32–39.
58. Kennedy, D., (2000). Database design and the reality of normalization. In: *Proceedings of the 13th Annual NACCQ Conference*.
59. Khan, W., Kumar, T., Zhang, C., Raj, K., Roy, A. M., & Luo, B., (2023). SQL and NoSQL database software architecture performance analysis and assessments—A systematic literature review. *Big Data and Cognitive Computing, 7*(2), 97.
60. Köhler, H., & Link, S., (2016). SQL schema design: Foundations, normal forms, and normalization. In: *Proceedings of the 2016 International Conference on Management of Data* (pp. 267–279).
61. Koltsidas, I., & Viglas, S. D., (2011). Data management over flash memory. In: *Proceedings of the 2011 ACM SIGMOD International Conference on Management of Data* (pp. 1209–1212).
62. Koonce, D., Huang, L., & Judd, R., (1998). EQL an express query language. *Computers & Industrial Engineering, 35*(1, 2), 271–274.
63. Kulkarni, K., & Michels, J. E., (2012). Temporal features in SQL: 2011. *ACM SIGMOD Record, 41*(3), 34–43.
64. Kumar, K. S., & Mohanavalli, S., (2017). A performance comparison of document-oriented NoSQL databases. In: *2017 International Conference on Computer, Communication and Signal Processing (ICCCSP)* (pp. 1–6). IEEE.
65. Kunda, D., & Phiri, H., (2017). A comparative study of NoSQL and relational database. *Zambia ICT Journal, 1*(1), 1–4.
66. Kung, H. J., & Tung, H. L., (2006). *A Web-Based Tool to Enhance Teaching/Learning Database Normalization, 1, 4-9.*
67. Kung, H. J., & Tung, H. L., (2006). An alternative approach to teaching database normalization: A simple algorithm and an interactive e-learning tool. *Journal of Information Systems Education, 17*(3).
68. Lakshmanan, L. V., Sadri, F., & Subramanian, S. N., (1999). On efficiently implementing SchemaSQL on an SQL database system. In: *VLDB* (Vol. 99, pp. 471–182).
69. Larson, P. Å., Blanas, S., Diaconu, C., Freedman, C., Patel, J. M., & Zwilling, M., (2011). *High-Performance Concurrency*

Control Mechanisms for Main-Memory Databases. arXiv preprint arXiv:1201.0228.

70. Li, D., Han, L., & Ding, Y., (2010). SQL query optimization methods of relational database system. In: *2010 Second International Conference on Computer Engineering and Applications* (Vol. 1, pp. 557–560). IEEE.
71. Li, X., Bian, F., Zhang, H., Diot, C., Govindan, R., Hong, W. H., & Lannaccone, G., (2005). Advanced indexing techniques for wide-area network monitoring. In: *21st International Conference on Data Engineering Workshops (ICDEW'05)* (pp. 1184–1184). IEEE.
72. Liu, Z. H., Hammerschmidt, B., & McMahon, D., (2014). JSON data management: Supporting schema-less development in RDBMS. In: *Proceedings of the 2014 ACM SIGMOD International Conference on Management of Data* (pp. 1247–1258).
73. Lourenço, J. R., Cabral, B., Carreiro, P., Vieira, M., & Bernardino, J., (2015). Choosing the right NoSQL database for the job: A quality attribute evaluation. *Journal of Big Data, 2*(1), 1–26.
74. Mapanga, I., & Kadebu, P., (2013). Database management systems: A NoSQL analysis. International *Journal of Modern Communication Technologies & Research (IJMCTR), 1*, 12–18.
75. Mason, R. T., (2015). NoSQL databases and data modeling techniques for a document-oriented NoSQL database. In: *Proceedings of Informing Science & IT Education Conference (InSITE)* (Vol. 3, No. 4, pp. 259–268).
76. McClure, R. A., & Krüger, I. H., (2005). SQL DOM: Compile time checking of dynamic SQL statements. In: *Proceedings of the 27th International Conference on Software Engineering* (pp. 88–96).
77. Melnik, S., Raghavan, S., Yang, B., & Garcia-Molina, H., (2001). Building a distributed full-text index for the web. In: *Proceedings of the 10th International Conference on World Wide Web* (pp. 396–406).
78. Melton, J., Michels, J. E., Josifovski, V., Kulkarni, K., & Schwarz, P., (2002). SQL/MED: A status report. *ACM SIGMOD Record, 31*(3), 81–89.
79. Mishra, P., & Eich, M. H., (1992). Join processing in relational databases. *ACM Computing Surveys (CSUR), 24*(1), 63–113.
80. Mohamed, M. A., Altrafi, O. G., & Ismail, M. O., (2014). Relational vs. NoSQL databases: A survey. *International Journal of Computer and Information Technology, 3*(03), 598–601.

81. Moniruzzaman, A. B. M., & Hossain, S. A., (2013). *NoSQL Database: New Era of Databases for Big Data Analytics-Classification, Characteristics, and Comparison*. arXiv preprint arXiv:1307.0191.
82. Nayak, A., Poriya, A., & Poojary, D., (2013). Type of NOSQL databases and its comparison with relational databases. *International Journal of Applied Information Systems, 5*(4), 16–19.
83. Neumann, T., Mühlbauer, T., & Kemper, A., (2015). Fast serializable multi-version concurrency control for main-memory database systems. In: *Proceedings of the 2015 ACM SIGMOD International Conference on Management of Data* (pp. 677–689).
84. Olsen, D. H., (2000). Accounting database design and SQL implementation revisited. *Review of Business Information Systems (RBIS), 4*(2), 53–68.
85. Oussous, A., Benjelloun, F. Z., Lahcen, A. A., & Belfkih, S., (2013). Comparison and classification of NoSQL databases for big data. *International Journal of Database Theory and Application, 6*(4.2013).
86. Pandey, V., Kipf, A., Neumann, T., & Kemper, A., (2018). How good are modern spatial analytics systems? *Proceedings of the VLDB Endowment, 11*(11), 1661–1673.
87. Parsons, R. G., Dale, A. G., & Yurkanan, C. V., (1974). Data manipulation language requirements for database management systems. *The Computer Journal, 17*(2), 99–103.
88. Patil, M. M., Hanni, A., Tejeshwar, C. H., & Patil, P., (2017). A qualitative analysis of the performance of MongoDB vs MySQL database based on insertion and retrieval operations using a web/android application to explore load balancing—Sharding in MongoDB and its advantages. In: *2017 International Conference on I-SMAC (IoT in Social, Mobile, Analytics and Cloud) (I-SMAC)* (pp. 325–330). IEEE.
89. Pokorny, J., (2011). NoSQL databases: A step to database scalability in web environment. In: *Proceedings of the 13th International Conference on Information Integration and Web-based Applications and Services* (pp. 278–283).
90. Provost, F., & Fawcett, T., (2013). Data science and its relationship to big data and data-driven decision making. *Big Data, 1*(1), 51–59.
91. Ramirez, J. A., (1973). *Automatic Generation of Data Conversions-Programs Using a Data Description Language (DDL)* (Vol. 1, 2). Technical Reports (CIS), 765.

92. Ramirez, J. A., Rin, N. A., & Prywes, N. S., (1974). Automatic generation of data conversion programs using a data description language. In: *Proceedings of the 1974 ACM SIGFIDET (now SIGMOD) Workshop on Data Description, Access, and Control* (pp. 207–225).
93. Raut, A. B. P. D. (2017). NOSQL database and its comparison with RDBMS. *International Journal of Computational Intelligence Research, 13*(7), 1645-1651. (2017). NOSQL database and its comparison with RDBMS. *International Journal of Computational Intelligence Research, 13*(7), 1645–1651.
94. Reddy, H. B. S., Reddy, R. R. S., Jonnalagadda, R., Singh, P., & Gogineni, A., (2022). Analysis of the unexplored security issues common to all types of NoSQL databases. *Asian Journal of Research in Computer Science, 14*(1), 1–12.
95. Rivero, C., & Ruiz, D., (2002). An advanced query capability description framework implementation. In: *International Conference on Enterprise Information System* (pp. 1–8). IEEE.
96. Rys, M., (2005). XML and relational database management systems: Inside Microsoft® SQL Server™ 2005. In: *Proceedings of the 2005 ACM SIGMOD International Conference on Management of Data* (pp. 958–962).
97. Sahatqija, K., Ajdari, J., Zenuni, X., Raufi, B., & Ismaili, F., (2018). Comparison between relational and NOSQL databases. In: *2018 41st International Convention on Information and Communication Technology, Electronics, and Microelectronics (MIPRO)* (pp. 0216–0221). IEEE.
98. Sahoo, S. S., Halb, W., Hellmann, S., Idehen, K., Thibodeau, Jr. T., Auer, S., & Ezzat, A., (2009). A survey of current approaches for mapping of relational databases to RDF (Vol. 1, pp. 113–130). W3C RDB2RDF Incubator Group Report.
99. Sallans, A., & Lake, S., (2014). Data management assessment and planning tools. *Research Data Management: Practical Strategies for Information Professionals,* 87–107.
100. Sanders, G. L., & Shin, S., (2001). Denormalization effects on performance of RDBMS. In: *Proceedings of the 34th Annual Hawaii International Conference on System Sciences* (p. 9). IEEE.
101. Santos, M. Y., & Costa, C., (2016). Data models in NoSQL databases for big data contexts. In: *Data Mining and Big Data: First International*

Conference, DMBD 2016, Bali, Indonesia, June 25–30, 2016; Proceedings 1 (pp. 475–485). Springer International Publishing.

102. Sareen, P., & Kumar, P., (2015). NoSQL database and its comparison with SQL database. *International Journal of Computer Science & Communication Networks, 5*(5), 293–298.
103. Scherzinger, S., & Sidortschuck, S., (2020). An empirical study on the design and evolution of NoSQL database schemas. In: *Conceptual Modeling: 39th International Conference, ER 2020, Vienna, Austria, November 3–6, 2020, Proceedings 39* (pp. 441–455). Springer International Publishing.
104. Schiefer, K. B., & Valentin, G., (1999). DB2 universal database performance tuning. *IEEE Data Eng. Bull., 22*(2), 12–19.
105. Schnaitter, K., Abiteboul, S., Milo, T., & Polyzotis, N., (2006). Colt: Continuous on-line tuning. In: *Proceedings of the 2006 ACM SIGMOD International Conference on Management of Data* (pp. 793–795).
106. Seshadri, P., & Paskin, M., (1997). PREDATOR: An OR-DBMS with enhanced data types. *ACM SIGMOD Record, 26*(2), 568–571.
107. Setiyadi, D., (2021). Database system development life cycle (DSDLC) on system libraries for data manipulation language (DML) using SQL server 2008:-. *Jurnal Mantik, 5*(2), 1065–1071.
108. Sharma, V., & Dave, M., (2012). SQL and NoSQL databases. *International Journal of Advanced Research in Computer Science and Software Engineering, 2*(8).
109. Shin, S. K., & Sanders, G. L., (2006). Denormalization strategies for data retrieval from data warehouses. *Decision Support Systems, 42*(1), 267–282.
110. Shoval, P., & Even-Chaime, M., (1987). Database schema design: An experimental comparison between normalization and information analysis. *ACM SIGMIS Database: The Database for Advances in Information Systems, 18*(3), 30–39.
111. Sidi, E., El, M., & Amin, E., (2016). Star schema advantages on data warehouse: Using bitmap index and partitioned fact tables. *International Journal of Computer Applications, 134*(13), 11–13.
112. Swaminathan, S. N., & Elmasri, R., (2016). Quantitative analysis of scalable NoSQL databases. In: *2016 IEEE International Congress on Big Data (BigData Congress)* (pp. 323–326). IEEE.

113. Tang, E., & Fan, Y., (2016). Performance comparison between five NoSQL databases. In: *2016 7th International Conference on Cloud Computing and Big Data (CCBD)* (pp. 105–109). IEEE.

114. Tang, F., You, I., Yu, S., Wang, C. L., Guo, M., & Liu, W., (2012). An efficient deadlock prevention approach for service-oriented transaction processing. *Computers & Mathematics with Applications, 63*(2), 458–468.

115. Teorey, T. J., Yang, D., & Fry, J. P., (1986). A logical design methodology for relational databases using the extended entity-relationship model. *ACM Computing Surveys (CSUR), 18*(2), 197–222.

116. Tiffany, R., (2003). Data manipulation language. In: *SQL Server CE Database Development with the. NET Compact Framework* (pp. 139–179). Berkeley, CA: A Press.

117. Truica, C. O., Boicea, A., & Trifan, I., (2013). CRUD operations in MongoDB. In: *2013 International Conference on Advanced Computer Science and Electronics Information (ICACSEI 2013)* (pp. 347–350). Atlantis Press.

118. Van, D. B. H., Van, D. L. R., & Visser, J., (2007). Quality assessment for embedded SQL. In: *Seventh IEEE International Working Conference on Source Code Analysis and Manipulation (SCAM 2007)* (pp. 163–170). IEEE.

119. Vera, H., Boaventura, W., Holanda, M., Guimaraes, V., & Hondo, F., (2015). Data modeling for NoSQL document-oriented databases. In: *CEUR Workshop Proceedings* (Vol. 1478, pp. 129–135).

120. Vera-Olivera, H., Guo, R., Huacarpuma, R. C., Da Silva, A. P. B., Mariano, A. M., & Holanda, M., (2021). Data modeling and NoSQL databases-a systematic mapping review. *ACM Computing Surveys (CSUR), 54*(6), 1–26.

121. Vyverman, M., De Baets, B., Fack, V., & Dawyndt, P., (2012). Prospects and limitations of full-text index structures in genome analysis. *Nucleic Acids Research, 40*(15), 6993–7015.

122. Wells, G., & Wells, G., (2001). DML: Inserting, updating, and deleting data. *Code Centric: T-SQL Programming with Stored Procedures and Triggers*, 133–148.

123. Woods, L., István, Z., & Alonso, G., (2014). Ibex: An intelligent storage engine with support for advanced SQL offloading. *Proceedings of the VLDB Endowment, 7*(11), 963–974.

124. Wu, K., Otoo, E. J., & Shoshani, A., (2001). A performance comparison of bitmap indexes. In: *Proceedings of the Tenth International Conference on Information and Knowledge Management* (pp. 559–561).

125. Wu, K., Stockinger, K., & Shoshani, A., (2008). Breaking the curse of cardinality on bitmap indexes. In: *Scientific and Statistical Database Management: 20th International Conference, SSDBM 2008, Hong Kong, China, July 9–11, 2008 Proceedings 20* (pp. 348–365). Springer Berlin Heidelberg.

126. Wu, M. C., & Buchmann, A. P., (1998). Encoded bitmap indexing for data warehouses. In: *Proceedings 14th International Conference on Data Engineering* (pp. 220–230). IEEE.

127. Xie, C., Su, C., Littley, C., Alvisi, L., Kapritsos, M., & Wang, Y., (2015). High-performance ACID via modular concurrency control. In: *Proceedings of the 25th Symposium on Operating Systems Principles* (pp. 279–294).

128. Xie, C., Su, C., Littley, C., Alvisi, L., Kapritsos, M., & Wang, Y., (2015). High-performance ACID via modular concurrency control. In: *Proceedings of the 25th Symposium on Operating Systems Principles* (pp. 279–294).

129. Yalamanchi, A., Srinivasan, J., & Gawlick, D., (2003). Managing expressions as data in relational database systems. In: *CIDR*.

130. Yang, X., Procopiuc, C. M., & Srivastava, D., (2009). Summarizing relational databases. *Proceedings of the VLDB Endowment, 2*(1), 634–645.

131. Zaki, A. K., (2014). NoSQL databases: New millennium database for big data, big users, cloud computing and its security challenges. *International Journal of Research in Engineering and Technology (IJRET), 3*(15), 403–409.

132. Zhang, X., Song, W., & Liu, L., (2014). An implementation approach to store GIS spatial data on NoSQL database. In: *2014 22nd International Conference on Geoinformatics* (pp. 1–5). IEEE.

133. Zhou, J., Bruno, N., & Lin, W., (2012). Advanced partitioning techniques for massively distributed computation. In: *Proceedings of the 2012 ACM SIGMOD International Conference on Management of Data* (pp. 13–24).

134. Zhu, Q., Gong, J., & Zhang, Y., (2007). An efficient 3D R-tree spatial index method for virtual geographic environments. *ISPRS Journal of Photogrammetry and Remote Sensing, 62*(3), 217–224.

CHAPTER

3

Relational Data Model and Relational Database Constraints

CONTENT

3.1. INTRODUCTION

A relational data model constitutes a fundamental database management system (DBMS) concept. Utilizing a relational database offers a systematic and methodical approach to storing and administrating data. Edgar F. Codd introduced the relational model during the 1970s, subsequently gaining widespread adoption as the predominant data model in the industry (Demuth & Hussmann, 1999) (Figure 3.1).

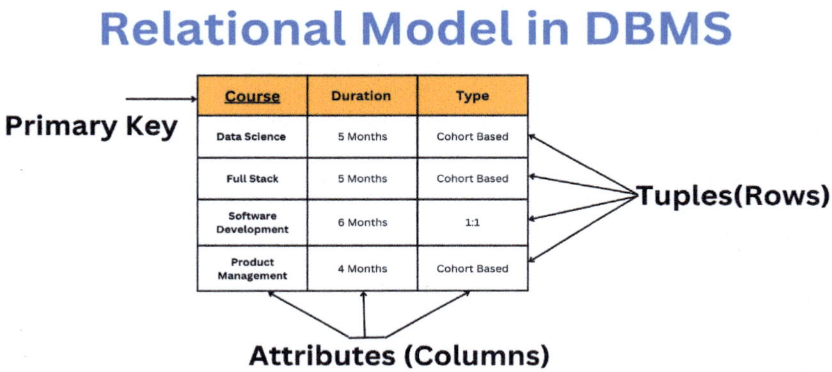

Figure 3.1. Illustration of relational model in DBMS.

Source: Board infinity, Creative Commons License.

Tables comprising rows and columns form the fundamental structure of relational data model. In context of tabular data, it can be observed that every row corresponds to a distinct entity or record. In contrast, each column pertains to an attribute or characteristic associated with that entity. As an illustration, within a database containing customer information, a table could contain various columns, including but not limited to "customer_id," "name," "email," and "phone_number" (Thapa & Giese, 2022).

The relational data model is advantageous because it can establish connections between tables. Establishing relationships within a table is facilitated using keys, which serve as distinctive identifiers. The relational model typically employs two distinct categories of keys: primary and foreign (Paredaens et al., 2012).

A fundamental concept in database design is the primary key, which refers to a singular column or a composite of columns that distinguishes each record within a given table. Implementing a unique identifier for each record in a table guarantees its distinctiveness and facilitates the effective retrieval and manipulation of data. As an illustration, the column denoted as "employee_id" may be assigned as the primary key within a tabular representation of staff members. Foreign keys establish inter-table relationships by referencing the primary key of another table. Referential integrity is enforced through their utilization, guaranteeing data consistency and accuracy in interrelated tables. In the context of relational databases, it is possible to establish a relationship between two tables, such as orders and customers, by utilizing a foreign key. Specifically, employee_id" column in orders table can be designated as foreign key, which references primary key "employee_id" in customers table (Delobel, 1978). A relational data model encompasses a collection of limitations and regulations, known as constraints, that govern the data housed within a database and the notion of keys. The implementation of constraints serves the purpose of upholding the integrity and consistency of data by obstructing the insertion of data that is deemed invalid or inconsistent. Various constraints are frequently employed in relational databases:

1. **Entity Integrity Constraint:** This restriction guarantees that the primary key attribute of a given table possesses a distinct and non-null value. The primary key attribute ensures the unique identification of each entity within a table and prohibits the absence or ambiguity of its corresponding value.
2. **Referential Integrity Constraint:** The enforcement of the relationship between tables is achieved using foreign keys as a constraint. The foreign key constraint guarantees that the values contained within the column designated as a foreign key in a given table correspond to values of a primary key in referenced table. The limitation mentioned above impedes the formation of isolated records and upholds the integrity of the data (Lin, 1988).
3. **Domain Constraint:** A domain constraint is a specification that establishes the permissible scope of values for a given column. The column's data type, length, format, and other attributes are specified by it. For instance, a column categorized as an "integer" might feature a domain constraint that restricts the permissible values to a particular range.
4. **Key Constraint:** A crucial limitation guarantees that the values in particular group or coloumn maintain uniqueness within given table. Implementing a mechanism that prohibits the insertion of redundant values and guarantees the fulfillment of the primary key constraint (He & Naughton, 2008).

5. **Check Constraint:** A check constraint is a logical expression that specifies a condition that must be satisfied by the data stored in a particular column or a group of columns. This feature facilitates the implementation of intricate data validation mechanisms that surpass the basic requirements of ensuring uniqueness or non-nullability. A check constraint can be employed to guarantee that a specific column representing an individual's age comprises values greater than zero and less than 150.
6. **Nullability Constraint:** This restriction dictates whether a given column can accommodate null values. In database management, a "not null" column is restricted from containing any missing or undefined values. In contrast, a column designated as "null" permits the non-existence of a value (Levene & Loizou, 2012).
7. **Integrity Constraint:** Integrity constraints refer to regulations that impose particular business rules or demands on the data. The database architect establishes the limitations in accordance with the particular requirements of the entity. An instance of an integrity constraint could entail that the remuneration allocated to an employee must not possess a negative value.

The imposition of these limitations by the relational model serves to safeguard the precision, coherence, and dependability of the information housed within a given database. In the event of a constraint violation, the DBMS will impede the operation that resulted in the violation, be it an insertion, update, or deletion of a record (Sumathi & Esakkirajan, 2007) (Figure 3.2).

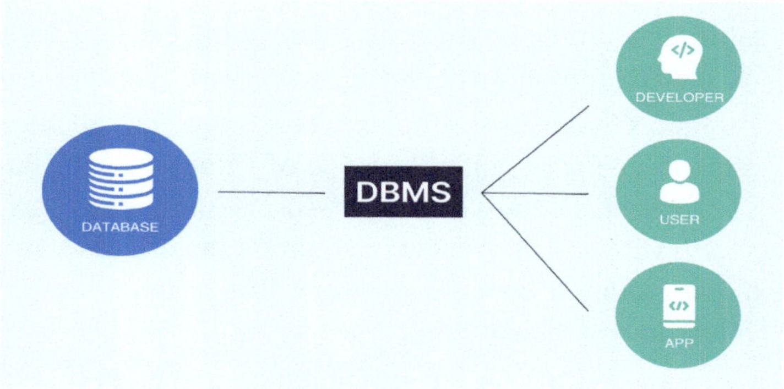

Figure 3.2. Schematic of database management system (DBMS).

Source: Margaret Rouse, Creative Commons License.

Relational database management systems offer means to establish and implement these limitations. Developers can define table constraints by utilizing Structured Query Language (SQL) statements. The DBMS automatically verifies and implements the constraints while performing data manipulation operations (Halpin & Morgan, 2010).

3.2. RELATIONAL MODEL CONCEPTS

The database is represented by a collection of relations in the relational model. In casual setting, it's noticeable that each relationship exhibits similarities to a value table or, to some extent, a record flat file. The term "flat file" is derived from the characteristic feature of its constituent records, which exhibit a simple and sequential format. The collection of records exhibits similarities to the depiction of the fundamental relational model. However, notable differences exist between relationships & files, we will observe soon (Malinowski & Zimányi, 2008).

When relation is conceptualized as tabular arrangement of values, each row within table denotes set of interrelated data values. The factual representation known as a row in a database often correlates to an object or connection in the real world. Terminology of table and its constituent columns facilitates the comprehension of the signification of the data contained in each row. The nomenclature of the columns, namely Name, Class, Student_number, & Major, delineate how data values in each row are to be comprehended, contingent upon the specific column in which the value is situated. Each element within a given column possesses an identical data type. In formal relational model context, tuple is term used to refer to a row, while an attribute is a term used to describe a column header (Kappel et al., 2001).

Additionally, a relation is the term used to denote a table. The domain of possible values represents data type that characterizes types of values which may be present in each column. The following terms, namely domain, tuple, attribute, & relation, shall be defined formally (Galan, 2020).

3.2.1. Domains, Relations, Tuples, and Attributes

In computer science, domain D refers to collection of fundamental and indivisible values. The term "atomic" refers to the indivisibility of each value within the domain in the context of the formal relational model. A prevalent approach to defining a domain involves specifying a particular data type from where constituent domain's data values are derived. Assigning name to domain can be advantageous as it aids in comprehending its values. The following are domains examples (Date, 1982) (Figure 3.3):

1 **Usa_phone_numbers:** A collection of 10 digit telephone numbers deemed legitimate within jurisdiction of United States.

2 **Local_phone_numbers:** This refers to collection of telephone numbers comprising seven digits that are deemed valid within a specific geographic region code in USA. The employment of local area phone numbers is swiftly becoming obsolete and being supplanted by traditional 10-digit phone numbers.

3 **Social_security_numbers:** A collection of legitimate Social Security Numbers (SSNs) consisting of 9 digits. SSN is distinctive alphanumeric identifier allocated to individuals in United States for employment, taxation, & benefits (Schek & Scholl, 1986).

Figure 3.3. Image of social security number (SSN).

Source: citizen path, Creative Commons License.

4 **Names:** A collection of character sequences that denote the appellations of individuals.

5 **Grade_point_averages:** Feasible numerical outcomes for calculated grade point averages are restricted to real numbers within the range of 0 to 4.

6 **Employee_ages:** The permissible range for the age of employees in a company is limited to integer values falling within the interval of 15 to 80.

7 **Academic_department_names:** A assemblage of nomenclatures for academic departments within a university, encompassing disciplines including Economics, Computer Science,& Physics (Chen, 1976).

8 **Academic_department_codes:** This is a compilation of academic department codes, encompassing designations such as 'CS,' 'ECON,' and 'PHYS."

Those mentioned above are referred to as logical definitions of domains. Each domain is associated with a specified data type or format. For example, the data type for the domain Usa_phone_numbers can be declared as a character string of the form *(ddd)ddd- dddd*, where each *d* is a numeric (decimal) digit and the first three digits form a valid telephone area code. Employee_ages variable is classified as an integer data type with values between 15 and 80. The data type assigned to Academic_department_names comprises a collection of character strings which accurately represent legitimate department names. Domain is assigned nomenclature, a specific data category, and a structure. Supplementary details about the interpretation of domain values may be provided. For instance, a numerical domain such as Person_weights ought to include corresponding units of measurement, such as kgs or lbs (Gadia, 1988). The structure of a relation schema, denoted as R(A1, A2, …, An), comprises a name for the relation, R, & set of attributes, A1, A2, …, An. In the relation schema R, each attribute Ai denotes a role fulfilled by a domain D. The domain of Ai is referred to as D & is symbolized by dom(Ai). A relation schema is a formal representation used to describe the structure of relation. The relation itself is identified by the name R. Numerical value denoting the number of attributes, represented by the symbol 'n,' within a relation schema is commonly referred to as the degree or arity of a relation. A septenary relation about the domain of university students would encompass seven distinct attributes that delineate the characteristics of each student. The following text is presented below (Codd, 1979):

STUDENT(Name: string, Ssn: string, Home_phone: string, Address: string, Office_phone: string, Age: integer, Gpa: real)

The STUDENT entity represents a subject in the relational model and consists of seven attributes. Each attribute can be assigned a specific data type:

- Name: string
- Ssn: string

- Home_phone: string
- Address: string
- Office_phone: string
- Age: integer
- Gpa: real

To define the domains for each attribute, we can map them to predefined types:

- dom(Name) = Names
- dom(Ssn) = Social_security_numbers
- dom(Home_phone) = USA_phone_numbers3
- dom(Address) = Addresses
- dom(Office_phone) = USA_phone_numbers
- dom(Age) = Integers
- dom(Gpa) = Grade_point_averages (with greater precision)

Based on the description, the attributes within the STUDENT relation can be identified by their position (Clifford & Croker, 1987):

- Name is the 1st attribute of the STUDENT relation.
- Ssn is the 2nd attribute of STUDENT relation.
- Home_phone is 3rd attribute of the STUDENT relation.
- Address is 4th attribute of the STUDENT relation.
- Office_phone is fifth attribute of STUDENT relation.
- Age is the sixth attribute of the STUDENT relation.
- Gpa is the seventh attribute of STUDENT relation.

This provides a clearer understanding of the characteristics and placement of the attributes within the STUDENT relation schema (Tansel, 1986).

A set of n-tuples defined by the symbol r(R), where R(A1, A2, …, An) is relation schema, is referred to as a "relation" (or "relation state"). This set is represented as r = {t1, t2, …, tm}. Every n-tuple, denoted as t, represents a sequentially arranged list of n values t =<v1, v2, …, vn>. Each value vi, where *vi,* 1 £ *i* £ *n*, is a member of dom (Ai) or is a unique NULL value. i-th element within the tuple t, denoting attribute Ai, is commonly denoted as t[Ai] or t.Ai (alternatively, t[i] if positional notation is employed). A concepts of relation intention and relation extension are frequently employed in schema R and relation state r(R), respectively (Codd, 1971).

Every tuple within relation denotes distinct student entity or object. Relation is represented in tabular form, wherein every tuple is represented as row, & every attribute is denoted by column header that signifies the function or meaning of values contained in particular column. The presence of NULL values in a STUDENT tuple indicates that certain attributes either have an unknown value or don't exist for that specific person.

Relation definition can be expressed more rigorously by utilizing concepts from set theory, as follows.

The concept of a relation, denoted as r(R), is a mathematical construct that represents a relation state. It corresponds to a relation with a degree of n, defined on domains A1, A2, …, An. This relation is essentially a subset of the Cartesian product (represented by the symbol '×') of the domains that define R:

r(*R*) Í (dom(*A*1) ′ dom(*A*2) ′ & ′ dom(*An*))

Cartesian product comprehensively delineates the complete set of feasible permutations of values derived from the respective underlying domains. The total number of tuples in the Cartesian product can be calculated by multiplying the cardinalities, denoted as |D|, of each individual domain D within a particular relation. It is important to note that this calculation assumes that all domains are finite.

By taking the Cartesian product of domains A1, A2, …, An, the resulting set will consist of tuples where each element of the tuple is selected from one of the domains. The number of tuples in the Cartesian product can be determined by multiplying the cardinalities of each domain (Li et al., 2005):

Total number of tuples = |A1| × |A2| ×. × |An|

This calculation provides an estimation of the total number of possible tuples within the Cartesian product, representing the entire set of combinations of values from the defined domains.

|dom(*A*1)| ′ |dom(*A*2)| ′ & ′ |dom(*An*)|

The cardinality product of all domains denotes maximum number of potential tuples or instances which could conceivably exist within any given state of a relation r(R). The current relation state, among various feasible combinations, exclusively represents the valid tuples corresponding to a specific state of actual world. Generally speaking, state of real world undergoes changes that result in a transformation of the relational state. The schema R is characterized by high stability and infrequent modifications, such as including an attribute to represent novel information that was not initially stored in the relation. Multiple attributes can share a common domain. The nomenclature assigned to attributes denotes distinct functions or connotations within the domain. In STUDENT entity, domain USA_phone_numbers serves as both Home_phone and Office_phone roles, denoting the respective phone numbers for a student's home and office. An additional attribute, which has not been illustrated, but shares the same domain, is the Mobile_phone attribute (Codd, 1972).

3.2.2. Characteristics of Relations

An antecedent relations definition encompasses distinct attributes which differentiate a relation from a table or a file. The following section will address several of characteristics mentioned above (Capello, 2002).

Did You Know?

Relations in a relational database exhibit the property of uniqueness, meaning each tuple within a relation is distinct and has a unique combination of attribute values.

- **Ordering of Tuples in a Relation:** A relation is defined by a set of tuples. From a mathematical standpoint, it is important to note that the elements within a set do not possess any inherent order. As a result, it follows that tuples contained within a relation lack any specific order. Stated differently, the ordering of tuples does not affect the sensitivity of a relation. In file storage, records are arranged in a specific order as they are physically stored on a disk or in memory. The sequence above denotes the file's initial, subsequent, i-th, and final entries. Likewise, the arrangement of rows is predetermined in presenting a relation as a table (Ng & Levene, 1997).

The absence of tuple ordering in a relation definition can be attributed to relations aiming to represent abstract facts. A single relation can have many tuple ordering specified. As an illustration, the tuples present in STUDENT relation have the potential to be arranged based on the values of attributes such as Name, Ssn, Age, or any other attribute. Relation definition is agnostic towards order, as it does not exhibit a bias towards any particular ordering. Thus, it can be inferred that the correlation is equivalent. In the context of data management, it is possible to specify a specific order for records of a file or rows of a table when representing a relation in the form of file or table (Tatarinov et al., 2002).

- **Value ordering in a Tuple & Different Definition of Relation:** As per definition of relation provided earlier, it can be inferred that an n-tuple denotes a sequence of n values arranged in a specific order. Therefore, it can be concluded that the order of values in tuple, & consequently the arrangement of attributes in a relation schema, holds significant importance. At higher level of abstraction, significance of attributes order & their respective values is relatively inconsequential, provided that the association between attributes & values remains intact (Selinger et al., 1979).

An alternative definition of a relation can be given, making the ordering of values in a tuple *unnecessary*. In this definition, a relation schema *R* = {*A*1, *A*2, …, *An*} is a *set* of attributes (instead of a list), and a relation state *r*(*R*) is a finite set of mappings *r* = {*t*1, *t*2, …, *tm*}, where each tuple *ti* is a mapping from *R* to *D*, and *D* is the union (denoted by È) of the attribute domains; that is, *D* = dom(*A*1) È dom(*A*2) È. È dom(*An*). In this definition, *t*[*Ai*] must be in dom(*Ai*) for 1 £ *i* £ *n* for each mapping *t* in *r*. Each mapping *ti* is called a tuple. (Wu et al., 2016). According to tuple definition as mapping, it can be considered grouping of pairs consisting of an attribute and its corresponding value. Each pair denotes the correspondence between a specific attribute Ai and its corresponding value vi from domain of Ai, denoted as dom(Ai). An impact of attribute ordering is inconsequential since the attribute identifier is consistently accompanied by its corresponding value (Table 3.1).

Table 3.1. Tabular Representation of Connection STUDENT with a Different Tuple Order.

STUDENT

Name	**Ssn**	**Home_phone**	**Address**	**Office_phone**	**Age**	**Gpa**
Dick Davidson	422–11–2320	NULL	3452 Elgin Road	(817)749–1253	25	3.53
Barbara Benson	533–69–1238	(817)839–8461	7384 Fontana Lane	NULL	19	3.25
Rohan Panchal	489–22–1100	(817)376–9821	265 Lark Lane	(817)749–6492	28	3.93
Chung-cha Kim	381–62–1245	(817)375–4409	125 Kirby Road	NULL	18	2.89
Benjamin Bayer	305–61–2435	(817)373–1616	2918 Bluebonnet Lane	NULL	19	3.21

Source: Codd, Creative Commons License.

According to this definition, it can be inferred that the two tuples exhibit complete similarity. This proposition is logical in a theoretical context, as there exists no rationale to prioritize the occurrence of one attribute value over another within tuple. The characteristics are organized as fields within a record in a predefined physical sequence when relation is implemented as file. Typically, primary definition of relation that we employ involves the imposition of order on attributes and values within tuples. This approach is favored due to its ability to streamline notation. Nevertheless, the alternate definition provided in this context is more comprehensive (Bhattacharya & Cortesi, 2009).

- **Values & NULLs in Tuples:** In the basic relational model, each element within tuple is considered an inseparable atomic value and cannot be further decomposed into smaller components. Therefore, using composite and multivalued attributes is prohibited in this context. The flat relational model is a nomenclature sometimes utilized to refer to this particular model. The foundational principles of the relational model were largely formulated with the underlying premise of 1st normal form assumption.6 Therefore, in fundamental relational model, it's necessary to depict multivalued attributes through distinct relations, while composite attributes are exclusively represented by component attributes (Zaniolo, 1982).

NULL values are crucial in database management systems as they denote attribute values that may be indeterminate or inapplicable to a given tuple. In such instances, a distinct value known as NULL is employed. Certain STUDENT tuples exhibit NULL values for their office phone attribute, which can be attributed to the absence of office space for these students, rendering the office phone attribute inapplicable. The student in question may have a missing value for their home phone, which could indicate either the absence of a home phone or a lack of knowledge regarding its existence. Typically, multiple interpretations exist for NULL values, including but not limited to an unknown value, an existing value that is inaccessible, or an attribute which is inapplicable to the given tuple (alternatively referred to as undefined value). An instance of the final category of NULL arises when introducing a Visa status attribute to the STUDENT entity, which exclusively pertains to tuples denoting international students. Formulating distinct codes to represent various interpretations of NULL values is feasible. Including various forms of NULL values in relational model operations has presented challenges & is beyond purview of our discourse (Biskup, 1984).

When a NULL value is subjected to arithmetic aggregations or comparisons with other values, its behavior is determined by a precise interpretation of the value. An instance of comparing two NULL values can result in ambiguities. In the scenario where both Customer A & B possess NULL addresses, it cannot be inferred that they share the same address. In designing a database, minimizing the usage of NULL values is advisable (Scannapieco & Batini, 2004).

- **Interpretation (Meaning) of a Relation**: A schema of a relation can be construed as declaration or form of assertion. The schema of STUDENT relation posits typical student entity possesses attributes such as Name, Ssn, Home_phone, Address, Office_ phone, Age, & Gpa. Each tuple within relation can be construed as factual statement or specific occurrence of assertion. The initial tuple affirms the existence of a STUDENT entity, denoted by the Name attribute "Benjamin Bayer," the Ssn attribute "305–61–2435," the Age attribute "19," & so forth (Wand et al., 1999).

It should be noted that certain relations may pertain to information regarding entities, while others may pertain to information regarding relationships. An illustration of a relation schema is MAJORS (Student_ ssn, Department_code), affirming that students pursue academic disciplines as their majors (Dey & Sarkar, 1996). The relation contains a tuple that establishes a correlation between students and their corresponding major discipline. Thus, relational model uniformly presents information about entities & relationships by utilizing relations. At times, the comprehensibility of a given relation may be affected when there is uncertainty about whether it represents an entity or a relationship type. This ambiguity can make it challenging to understand the nature and purpose of the relation (Zvieli & Chen, 1986). A possible alternative perspective on a relation schema is to view it as a predicate. Under this framework, the values contained within each tuple are construed as values that fulfill the predicate. Predicate denoted as STUDENT (Name, Ssn.) holds for the set of five tuples in the STUDENT relation. The tuples above denote five distinct propositions or factual statements about the physical world. The interpretation mentioned above holds significant utility in the domain of logical programming languages, particularly in Prolog, as it facilitates the integration of the relational model into these languages. The closed world assumption is a postulate that posits the veracity of solely those facts contained within the extension or state of the relation(s) in the universe. Any alternative combination of values renders the predicate invalid (Shenoi & Melton, 1989).

3.2.3. Relational Model Notation

The following notation will be used in our presentation:

1. The notation R(A1, A2, …, An) represents a relation schema R of degree n.
2. An uppercase letters Q, R, and S denote the relation names.
3. A lowercase letters q, r, and s represent states of relation.
4. Symbols t, u, and v are used to represent tuples.
5. Typically, nomenclature of relation schema, including STUDENT, denotes present collection of tuples within that relation, known as the current relation state. Conversely, STUDENT(Name, Ssn.) pertains solely to relation schema (Marcos et al., 2003).
6. The dot notation R.A can qualify an attribute A with its corresponding relation name R, such as in the STUDENT case. Please provide your name and age as a student. This phenomenon occurs due to the potential for identical nomenclature to be assigned to two distinct attributes across disparate relations. Each attribute name within a given relation must be unique.
7. In relation r(R), an n-tuple t can be represented as t = <v1, v2., vn>, where each vi corresponds to the value of attribute Ai. The notation presented pertains to values of components within tuples:
8. Notations t[Ai], t.Ai, & occasionally t[i] are used interchangeably to denote value vi in t corresponding to the attribute (Getoor & Sahami, 1999).
9. The subtype of values <vu, vw., vz> from t, corresponding to attributes listed in R as Au, Aw., Az, can be denoted as either t[Au, Aw., Az] or t.(Au, Aw., Az) (Liu et al., 2003).

An illustrative instance to contemplate is the tuple denoted by t, which comprises the following attributes: 'Barbara Benson' denoting the name, '533–69–1238' and '(817)839–8461' representing the phone numbers, '7384 Fontana Lane' indicating the address, NULL signifying the absence of a value for a particular attribute, 19 indicating the age, and 3.25 representing a numerical attribute. Based on the STUDENT entity, it can be inferred that t[Name] corresponds to the value of <'Barbara Benson'>, while t[Ssn, Gpa, Age] corresponds to the values of <'533–69–1238,' 3.25, 19> (Pieris, 2013) (Figure 3.4).

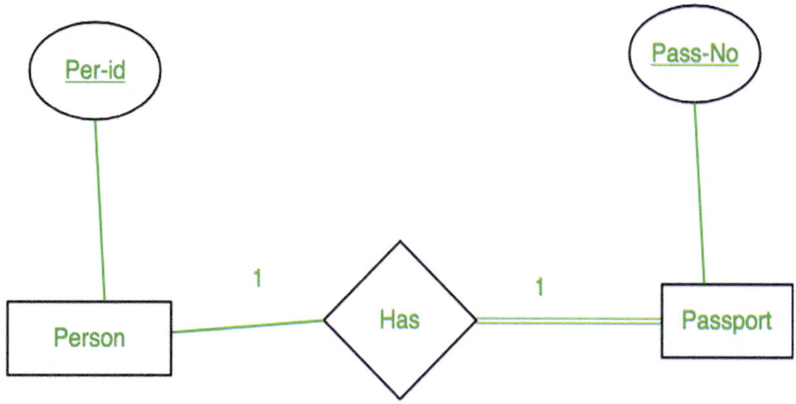

Figure 3.4. Image of example for relational model.

Source: Geeks for Geeks, Creative Commons License.

3.3. RELATIONAL MODEL CONSTRAINTS AND RELATIONAL DATABASE SCHEMAS

Up to this point, an analysis has been conducted on the characteristics of personal relationships. Within the realm of relational databases, it is a frequent occurrence to come across numerous relations that contain tuples showcasing a variety of forms of interdependence. At a particular point in time, the comprehensive condition of database will correspond to conditions of each its relations. Database states are commonly constrained by various limitations or restrictions governing their values. The limitations mentioned are inferred from the regulations inherent in the microcosm that the database represents (Fahrner & Vossen, 1995). The section pertains to delineating limitations imposed on data within a relational database, which can be expressed as constraints. The limitations imposed on databases can typically be classified into three primary groupings:

1 The limitations that are intrinsic to the data model. The constraints based on model itself or not explicitly stated are commonly referred to as inherent model-based or **implicit constraints**.

2 Limitations which are explicitly stated in schemas of data model, often by defining them in **Data Definition Language (DDL).** The constraints based on a schema or explicitly defined are called **schema-based constraints**.

3 Certain constraints are not directly expressible in schemas of a data model. Consequently, these constraints must be articulated & enforced by application programs. The constraints or rules based on applications, semantics, or business are called application-based, semantic, or business constraints (Markowitz & Makowsky, 1990).

The attributes of relationships pertain to the fundamental limitations of the relational paradigm and are classified under the primary category. An instance of an inherent constraint is the restriction which a relation cannot contain duplicate tuples. The constraints under consideration in this section belong to second category, which refers to constraints which can be articulated within schema of relational model through DDL. Limitations falling under the third category are broader, about both the semantics and conduct of attributes, & are arduous to articulate & implement within data model. As a result, they're typically verified by application programs responsible for executing database updates (Yan & Ling, 1993) (Figure 3.5).

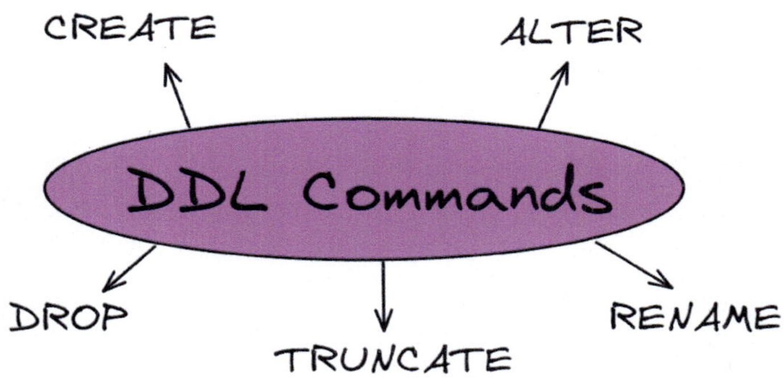

Figure 3.5. Illustration of DDL commands.

Source: Algo daily, Creative Commons License.

Data dependencies constitute a significant category of constraints, encompassing functional and multivalued dependencies. Relational database testing primarily involves assessing the efficacy of the design and is commonly executed through the normalization process. The constraints based on schema include but are not limited to domain constraints, constraints on NULLs, key constraints, referential integrity constraints, entity integrity constraints (Cerbah, 2008).

3.3.1. Domain Constraints

According to domain constraints, each attribute A in a tuple must have an atomic value that comes from domain dom(A). Domains are commonly associated with data types that encompass numeric values. These data types may include standard integer data types, including integer, short & long integer, as well as real number data types, such as float and double-precision float. In addition to standard data types, such as integers and floats, various other data types are also available in programming. These include characters, Booleans, fixed-length strings, & variable-length strings. Furthermore, specialized data types, such as time, timestamp, date & money, are also available in programming. It is plausible to depict additional domains as values subset from a given data type or as an enumerated data type that explicitly enlists each feasible values (Hui & Gray, 2000).

3.3.2. Key Constraints and Constraints on NULL Values

Formal relational model defines relation as tuples collection. As per the definition, it can be inferred that each element within a set is unique. Therefore, every tuples present in relation must also be unique. It implies that no 2 tuples will have an identical value amalgamation across all their attributes. Typically, relation schema R comprises additional attribute subsets that possess the characteristic whereby no 2 tuples within any given relation state r of R may share an identical combination of values for said attributes. Let SK be a subset of attributes in relation state R. There is a constraint like this for any two different tuples in R (Candan et al., 1997):

$t1[SK]\ ^{1}\ t2[SK]$

Relation schema R can be associated with a set of attributes SK, referred to as a super key. A superkey, denoted as SK, establishes a uniqueness constraint whereby it is impossible for two separate tuples within any given state r of R to possess identical values for SK (Lerat & Lipski, 1986). It is a commonly accepted fact in database management that every relation possesses a default superkey, essentially complete attributes set associated with it. A superkey may contain redundant attributes, whereas a key is a more valuable concept as it lacks redundancy. In the context of relation schema R, key K is considered superkey with added characteristic that removal of any attribute A from K results in a set of attributes K,' which no longer qualifies as a superkey of R. Therefore, key exhibits 2 distinct properties (Halpin & Ritson, 1992):

1 A fundamental principle in relational databases is that no two tuples within a given state of a relation can possess identical values for the attributes that comprise key. The characteristic as mentioned above is also applicable to superkey.

2 The aforementioned is minimal superkey, meaning it's superkey that satisfies the uniqueness constraint in condition 1 and cannot have any attributes removed from it. The presence of this attribute does not constitute a necessity for a superkey.

While the initial characteristic pertains to keys and superkeys, the subsequent characteristic is solely necessary for keys (Raghunathan et al., 2001). Therefore, it can be stated that a key possesses the characteristic of being a superkey, but the reverse is not necessarily true. STUDENT relation. Set of attributes {Ssn} is considered primary key of the entity STUDENT, as it ensures that there cannot exist two tuples within the STUDENT entity with identical values for the attribute Ssn.8 A superkey is any collection of attributes that encompasses Ssn, such as {Ssn, Name, Age}. Superkey {Ssn, Name, Age} does not qualify as key for the entity STUDENT since removing either Name or Age or both from the set as mentioned above would still result in a superkey. Any superkey that is constituted by a solitary attribute is concurrently key. For key to possess the uniqueness property, it is imperative that all of its attributes collectively satisfy this requirement in a multi-attribute scenario (Atzeni & Morfuni, 1986).

Tip

When defining key constraints in a relational database, choose attributes or combinations of attributes that uniquely identify each tuple, ensuring data integrity .and preventing duplicates

The identification of each tuple within a relationship can be achieved by utilizing a key attribute that possesses a unique value. The SSN value of 305–61–2435 serves as a unique identifier for the tuple that pertains to Benjamin Bayer within the STUDENT relation. It should be noted that a collection of characteristics that make up a key is an inherent feature of the relation schema. This represents a limitation that must be satisfied by all legitimate relation states within the schema. The determination of a key is based on the semantic interpretation of the attributes, and the property of being time-invariant is inherent. The integrity constraint must remain valid even upon the insertion of additional tuples into the relation. The designation of the Name attribute of the STUDENT relation as a key is not feasible nor advisable due to the potential existence of two students with identical names in a valid state (Eckert et al., 2020).

It is common for a relational schema to possess multiple keys. In this instance, every individual key is called a candidate key. The CAR relationship possesses two potential keys, namely License_number & Engine_serial_number. It's customary to assign one of candidate keys as primary key for given relation. The candidate key is a set of values that serves the purpose of identifying tuples within the relation. The customary practice is to underline attributes which constitute primary key of a relation schema. It's crucial to remember that choosing a primary key can be arbitrary when a relational schema has several candidate keys. Nonetheless, it is generally advisable to opt for primary key that comprises a single attribute or a limited number of attributes. The candidate keys that remain are distinguished as being unique and not denoted by an underline.

One additional limitation on attributes pertains to the allowance or disallowance of NULL values. As an illustration, if each tuple about STUDENT must possess a legitimate and non-NULL value for Name, then the attribute of Name for STUDENT is restricted to NOT NULL (Goldstein, 1981).

3.3.3. Relational Databases and Relational Database Schemas

Definitions & limitations that've been previously discussed pertain to individual relations & their corresponding attributes. Typical relational database comprises multiple relations, wherein tuples within these relations are interconnected in diverse ways. This section defines relational database & relational database schema (Chiang et al., 1994).

Schema for a relational database. The set S comprises collection of relation schemas denoted as S = {R1, R2, ..., Rm}, along with a set of integrity constraints IC. The relation set states DB = {r1, r2, ..., rm} in a relational database state10 DB of S is defined such that each ri represents a state of Ri & fulfills integrity constraints specified in IC (Astrova, 2004). Relational database schema denoted as COMPANY = {EMPLOYEE, DEPARTMENT, DEPT_LOCATIONS, PROJECT, WORKS_ON, DEPENDENT}. The attributes that are underlined denote primary keys. (Nyulas et al., 2007).

A term "relational database" encompasses its schema and present state by default. In the realm of databases, a state that fails to adhere to all of the prescribed integrity constraints is deemed invalid. Conversely, a state of the database that conforms to all of the specified integrity constraints within the defined set of constraints, known as IC, is considered valid (Meier et al., 2008).

The Dnumber attribute found in both DEPARTMENT & DEPT_LOCATIONS entities represents the same real-world concept. A department is assigned a numerical identifier. The notion as mentioned above is referred to as Dno within the context of EMPLOYEE and as Dnum about PROJECT. The nomenclature of attributes which denote a common real-world concept may exhibit variance across distinct relations. In some cases, dissimilar concepts may possess identical nomenclature across various relationships. An instance of redundancy in attribute naming could be observed where attribute Name is utilized for both the Pname of PROJECT & Dname of DEPARTMENT. It would result in 2 attributes with identical names but denote distinct real-world concepts, namely project and department names (McMinn et al., 2017).

Some initial versions of relational model were based on the assumption that a particular real-world concept would exhibit consistent naming conventions across all relationships when represented as an attribute. Utilizing a singular real-world concept in various capacities or connotations within a single relationship presents challenges.

The depicted EMPLOYEE relation exhibits duplication of the Social Security number concept, whereby it functions as both SSN of the employee & SSN of supervisor. It is imperative to utilize unique attribute names, such as Ssn and Super_ssn, to distinguish their intended meanings within a shared relation (Telnarova, 2010).

Every relational DBMS must possess a DDL that enables the definition of relational database schema. Relational database management systems predominantly employ SQL as their primary language for performing this function (Fonkam & Gray, 1992) (Table 3.2).

Table 3.2. Tabular Representation of One Possible Database State for the COMPANY Relational Database Schema.

EMPLOYEE

Fname	**Minit**	**Lname**	**Ssn**	**Bdate**	**Address**	**Sex**	**Salary**	**Super_ssn**	**Dno**
John	B	Smith	123456789	1965–01–09	731 Fondren, Houston, TX	M	30000	333445555	5
Franklin	T	Wong	333445555	1955–12–08	638 Voss, Houston, TX	M	40000	888665555	5

Alicia	J	Zelaya	999887777	1968–01–19	3321 Castle, Spring, TX	F	25000	987654321	4
Jennifer	S	Wallace	987654321	1941–06–20	291 Berry, Bellaire, TX	F	43000	888665555	4
Ramesh	K	Narayan	666884444	1962–09–15	975 Fire Oak, Humble, TX	M	38000	333445555	5
Joyce	A	English	453453453	1972–07–31	5631 Rice, Houston, TX	F	25000	333445555	5
Ahmad	V	Jabbar	987987987	1969–03–29	980 Dallas, Houston, TX	M	25000	987654321	4
James	E	Borg	888665555	1937–11–10	450 Stone, Houston, TX	M	55000	NULL	1

DEPARTMENT

Dname	Dnumber	Mgr_ssn	Mgr_start_date
Research	5	333445555	1988–05–22
Administration	4	987654321	1995–01–01
Headquarters	1	888665555	1981–06–19

DEPT LOCATION

Dnumber	Dlocation
5	Houston
1	Stafford
4	Bellaire
5	Sugarland
5	Houston

Source: Shashank Tiwari, Creative Commons License.

Integrity constraints are formally defined and imposed on a database schema, with the expectation that they will remain valid for all legitimate states of the database that adhere to that schema.

Apart from the domain, key, & NOT NULL constraints, the relational model encompasses two additional types of constraints, namely entity integrity & referential integrity (Cunha et al., 2009).

3.3.4. Integrity, Referential Integrity, and Foreign Keys

Principle of entity integrity constraint posits which it is impermissible for any primary key value to assume a NULL state. The rationale behind this is primary key's value serves as a unique identifier for distinct tuples within a given relation. NULL values in the primary key indicate the inability to identify certain tuples. If multiple tuples possess NULL values for their primary keys, it may not be feasible to differentiate between them when attempting to reference them from other relations (Calero et al., 2001).

Individual relations are subject to the specification of key and entity integrity constraints. Referential integrity constraint pertains to the specification between two relations upholding the coherence of tuples within said relations. Referential integrity constraint is an informal principle stipulating that a tuple within one relation that references another must reference a pre-existing tuple within that relation. Attribute Dno of the EMPLOYEE entity provides department number corresponding to the department where each employee is employed. Therefore, the value of Dno in each tuple of EMPLOYEE must correspond to Dnumber value of a tuple in DEPARTMENT relation (Levene & Loizou, 2001).

To provide a formal definition of referential integrity, it is necessary first to establish concept of a foreign key. Following conditions pertain to foreign key and establish referential integrity constraint between 2 relational schemas, R1 & R2. In a relational schema, a collection of attributes is denoted as FK. Relation R1 is said to have a foreign key which references relation R2 if it adheres to prescribed rules (Meurice et al., 2014):

1 Attributes contained within foreign key (FK) possess identical domains as primary key (PK) attributes of R_2. As a result, attributes within FK are commonly referred to as referencing or referring to R_2 relation.

2 In current state of $r_1(R_1)$, tuple t_1 contains an FK value that is either a PK value for a tuple t2 in current state of $r_2(R_2)$ or it is NULL. When $t_1[FK]$ is equal to $t_2[PK]$, it can be stated that tuple t1 is referencing or referring to the tuple t_2 (Reinert, 1995).

The nomenclature employed in this definition designates R1 as referencing relation, while R2 is identified as referenced relation. Provided that these two criteria are met, it can be stated that referential integrity constraint exists between R1 and R2. In a database comprising numerous relations, it is common to encounter many referential integrity constraints.

To delineate these limitations, it is imperative to possess a lucid comprehension of significance or function that every attribute or group of attributes fulfills in diverse relation schemas of the database. Referential integrity constraints commonly emerge from the interconnections among the entities depicted by the relation schemas. An instance that can be considered is the database. The attribute Dno in the EMPLOYEE relation denotes the department where an employee is employed. Therefore, Dno is identified as a foreign key of EMPLOYEE that references DEPARTMENT relation. The aforementioned statement states that the value of Dno for any tuple t1 in the EMPLOYEE relation must coincide with a value of the Dnumber attribute in any tuple t2 of the DEPARTMENT relation, which is the main key for DEPARTMENT. Alternatively, the value of Dno may be NULL if the employee does not already belong to a department or will be assigned to one in the future. The tuple about the employee named 'John Smith' refers to the tuple associated with the 'Research' department, thereby signifying that 'John Smith' is employed by this particular department (Vidaković et al., 2016).

It should be noted that a foreign key can reference its relation. The attribute "Super_ssn" in the "EMPLOYEE" relation denotes the superior of an employee, who is also an employee and is represented by a tuple in the same "EMPLOYEE" relation.

Therefore, Super_ssn attribute is a foreign key that establishes a reference to the EMPLOYEE relation within itself. The tuple about employee 'John Smith' refers to the tuple of employee 'Franklin Wong,' thereby implying that 'Franklin Wong' holds the supervisor position for 'John Smith.' Referential integrity constraints can be visually represented through diagrams, wherein a directed arc is drawn from each foreign key to the relation it references. To enhance clarity, the arrowhead can indicate the primary key of the relation being referenced. The referential integrity constraints of the schema are presented in the format. To enforce integrity constraints on the database states, said constraints must be explicitly specified on the relational database schema as an integral part of its definition (Aleksić et al., 2011).

Therefore, DDL incorporates clauses for defining diverse categories of constraints, enabling the DBMS to enforce them automatically. Most relational database management systems support constraints related to entity integrity, key & referential integrity. The abovementioned limitations are explicitly outlined as component of the data DDL (Yue, 2022).

3.3.5. Other Types of Constraints

An antecedent integrity constraints have been incorporated into the DDL due to their ubiquitous presence in most database applications. Notwithstanding, relational databases do not encompass a broad category of overall limitations, denoted as semantic integrity constraints, that may necessitate explicit specification and implementation. Illustrations of such limitations include the provision that an employee's remuneration ought not to surpass that of their immediate superior and the ceiling on the total number of hours that an employee may devote to all projects in a given week being 56. Constraints may be defined and implemented in the database-updating application programs or by utilizing a constraint specification language that is not specific to any particular purpose. Triggers and assertions represent viable mechanisms that can be employed. Using CREATE ASSERTION & CREATE TRIGGER statements in SQL can serve this objective. In practice, application programs are frequently employed to verify constraints of this nature instead of utilizing constraint specification languages, which can prove challenging and intricate to navigate (Luković et al., 2007).

The functional dependency constraint is type of constraint that creates functional relationship between two sets of attributes, namely X & Y.

Functional dependency X not equal to Y is constraint which mandates the determination of a singular value of Y in every relation states based on value of X.

Constraints that have been previously discussed may be classified as state constraints, as they establish limitations which legitimate state of database must adhere to. Transition constraints can be established to address alterations in the database's state, constituting an additional category of constraints. One instance of a transition constraint is, "The remuneration of a staff member can solely experience an upward trend." The enforcement of such limitations is commonly carried out by application programs or designated through active rules and triggers (Maatuk et al., 2008).

3.4. UPDATE OPERATIONS, TRANSACTIONS AND DEAL WITH CONSTRAINT VIOLATIONS

Relational model encompasses two main operations categories: retrievals and updates. A new relation is created when algebraic operators are applied to a set of preexisting relations using a relational algebra expression. Primary purpose of this operation is to retrieve information from a database through querying. The user constructs a query that defines the target data and utilizes relational operators to extract said data, resulting in the formation of a novel relation. A resultant relation represents the outcome of the user's inquiry (Di Nitto & Tanca, 1996).

This section is focused on the operations related to modifying or updating a database. The fundamental actions that can alter the states of relations within a database are Insertion, Deletion, & Modification (or Update). The users can perform various operations on data records, such as inserting new data, deleting old data, or modifying existing data. The insertion operation, called "Insert," adds one or more new tuples to given relation. The deletion operation, "Delete," removes tuples from the relation. Lastly, the modification operation "Update" or "Modify" is utilized to alter values of certain attributes within existing tuples (Gehani & Jagadish, 1991). The application of these operations mustn't result in any violation of the integrity constraints that have been specified in the schema of the relational database. This section entails a discussion of the various constraints types which could potentially be breached by each operation & corresponding actions which may be implemented in response to such violations. The database is utilized for illustrative purposes, focusing exclusively on a key, entity, and referential integrity constraints. Examples are provided for every operation type, and any potential constraints each operation may violate are discussed (Boydens et al., 1998).

3.4.1. The Insert Operation

Inserting a new tuple t into a relation R involves providing a list of attribute values through the Insert operation. The act of inserting can potentially contravene any of the four categories of constraints that were previously expounded upon. The violation of domain constraints may occur when an attribute value is provided which either doesn't belong to corresponding domain or isn't of suitable data type. The violation of key constraints may occur when key value present in a new tuple t is already present in another tuple within relation r(R). Entity integrity violation may occur if any constituent of primary key of a newly introduced tuple, denoted as t, is rendered NULL. The violation of referential integrity may occur if a tuple referenced in a relationship does not exist, as indicated by value of a foreign key in t. The following examples are presented to elucidate the discourse mentioned above (Jajodia & Sandhu, 1991).

1. **Operation***:* Please execute the command to insert the tuple <'Cecilia,' 'F,' 'Kolonsky,' NULL, '1960–04–05,' '6357 Windy Lane, Katy, TX,' F, 28000, NULL, 4> into the EMPLOYEE relation. The insertion mentioned above is deemed invalid as it breaches entity integrity constraint, specifically primary key Ssn being assigned a NULL value, thus leading to its rejection.
2. **Operation:** The following SQL command is proposed for the insertion of a new record into the EMPLOYEE table: INSERT INTO EMPLOYEE (FIRST_NAME, MIDDLE_INITIAL, LAST_NAME, SSN, BIRTH_DATE, ADDRESS, GENDER, SALARY, SUPER_SSN, DEPT_NUM) VALUES ('Alicia,' 'J,' 'Zelaya,' '999887777,' '1960–04–05,' '6357 Windy Lane, Katy, TX,' 'F,' 28000, '987654321,' 4); The attempted insertion is deemed invalid due to a key constraint violation. This is because another tuple already exists with identical Ssn value in the EMPLOYEE relation. As a result, the insertion is rejected.
3. **Operation:** The following command is to be executed to insert the tuple <'Cecilia,' 'F,' 'Kolonsky,' '677678989,' '1960–04–05,' '6357 Windswept, Katy, TX,' F, 28000, '987654321,' 7> into the EMPLOYEE relation. The insertion mentioned above is in contravention of the referential integrity constraint that has been stipulated on Dno in the EMPLOYEE table. This is due to the absence of a corresponding tuple in DEPARTMENT table with a Dnumber value of 7, the referenced value.
4. **Operation:** The following SQL statement is proposed for new record insertion into EMPLOYEE table: INSERT INTO EMPLOYEE (First_name, last_name, gender, ssn, birth_date, address, marital_status, salary, supervisor_ssn, department_number) VALUES ('Cecilia,' 'F,' 'Kolonsky,' '677678989,' '1960–04–05,' '6357 Windy Lane, Katy, TX,' 'F,' 28000, NULL, 4). The insertion mentioned above meets each specified criteria and thus can be deemed permissible (Elmasri et al., 1985).

If an insertion contravenes one or multiple constraints, the standard course of action is to decline the insertion. It would be advantageous if the DBMS could furnish a rationale to the user regarding the cause of the insertion's rejection in this scenario. An alternative approach involves rectifying the cause of the rejection. However, this method is not commonly employed for violations arising from the Insert operation. Instead, it is frequently utilized to address Delete and Update operations violations. During the initial operation, DBMS may prompt user to input a value for Ssn & subsequently authorize insertion if legitimate Ssn value is supplied. During operation 3, the DBMS can prompt the user to modify the value of Dno to a valid value or set it to NULL. Alternatively, DBMS may request the user to insert a DEPARTMENT tuple with Dnumber equal to 7 before accepting the original insertion. If a user inserts a department 7 tuple with a Mgr_ssn that isn't in EMPLOYEE relation, it could cascade (Ben-Zvi, 1982).

3.4.2. Delete Operation

Deleting data can solely result in a breach of referential integrity. This phenomenon arises when the tuple that is being deleted is referred to by foreign keys from other tuples within database. To perform a deletion operation, criterion based on attributes of relation is utilized to identify tuple or tuples that are to be removed. The following are instances (Arqués & Michel, 1993).

1. **Operation:** Remove tuple with combination of Essn = '999887777' and Pno = 10 from the WORKS_ON relation. The deletion operation is deemed acceptable as it precisely removes a single tuple.
2. **Operation:** Remove the tuple with the attribute Ssn equal to '999887777' from the EMPLOYEE relation. The deletion cannot be deemed acceptable as tuples in the WORKS_ON relation which refer to the tuple mentioned above exist. Therefore, deleting the tuple in the EMPLOYEE table would lead to referential integrity violations.
3. **Operation:** Remove the tuple with the attribute Ssn equal to '333445555' from the EMPLOYEE relation. Removing this tuple is anticipated to exacerbate referential integrity violations, as tuples from various relations including EMPLOYEE, DEPARTMENT, WORKS_ON, and DEPENDENT, are currently referencing it (Adya et al., 2002).

There exist multiple alternatives in case a deletion operation results in a breach. The initial alternative, "restrict," involves declining the removal. The alternative approach, cascade, involves attempting to propagate deletion process by eliminating tuples which refer to deleted tuple. In the second operation, the database management system can automatically remove the problematic tuples from the WORKS_ON table where SSN of employee is '999887777.' The third alternative, "set null" or "set default," involves altering values of referencing attributes responsible for violation. Every value is either replaced with NULL or substituted with a different valid tuple that serves as the new reference. It should be noted that it is not possible to give a reference property in a primary key that results in a violation a NULL value. Doing so would result in a breach of entity integrity. It is also feasible to have various combinations of these three alternatives. To circumvent the occurrence of a violation resulting from operation 3, DBMS may execute an automatic deletion of all tuples from the tables WORKS_ON and DEPENDENT, wherein the attribute Essn is equal to '333445555.' The tuples present in the EMPLOYEE table that have the attribute Super_ssn equal to '333445555' & tuple in DEPARTMENT table with the attribute Mgr_ssn equal to '333445555' are subject to modification of their Mgr_ssn & Super_ssn values, which can be replaced by other valid values or set to NULL.

Automatically eliminating WORKS_ON and DEPENDENT tuples that are associated with an EMPLOYEE tuple may be a reasonable approach. However, it may not be reasonable to remove other EMPLOYEE tuples or a DEPARTMENT tuple.

When a referential integrity restriction is stated in DDL, it is often, DBMS enables database designer to designate appropriate option in the event of a constraint violation (Ai & Zhang, 2019).

3.4.3. The Update Operation

A given relation's tuples' attribute values can be changed through Update operation. To select tuple or tuples that require modification, and it's imperative to specify condition of relation attributes. Following are instances.

1 **Operation:** Modify the EMPLOYEE tuple with the Ssn value of '999887777' to reflect a salary 28000. The outcome of the situation is deemed satisfactory and meets the required standards.

2 **Operation:** Modify Dno attribute of EMPLOYEE tuple where the Ssn attribute is equal to '999887777' to have a value of 1. The outcome of the situation is deemed satisfactory and meets the required standards.

3 **Operation:** Modify Dno attribute of EMPLOYEE tuple where Ssn equals '999887777' to a value of 7. The outcome is deemed unsatisfactory as it breaches the principle of referential integrity

(Dayal & Bernstein, 1982).

4 **Operation:** Modify SSN attribute of EMPLOYEE tuple, specifically the one with an SSN value of '999887777,' to '987654321.' The outcome is deemed unacceptable due to violating the primary key constraint, wherein a value already existing as primary key in another tuple has been repeated. Furthermore, it violates referential integrity constraints since there are other relations which reference pre-existing values of SSN attribute (Baumann et al., 2005).

Modifying a non-primary key and non-foreign key attribute typically does not result in any complications. The DBMS must verify that the updated value matches the appropriate data type and domain. The process of altering primary key value resembles deleting tuple & subsequently inserting a new one in its stead, as the primary key identifies tuples. Therefore, the abovementioned concerns regarding (Insert) and (Delete) are relevant. In the event of a modification to a foreign key attribute, the DBMS must ensure that updated value pertains to an extant tuple within referenced relation or is alternatively assigned a NULL value. Analogous alternatives are available to address referential integrity breaches resulting from Update, akin to the solutions deliberated for the Delete function. In the context of database management, the establishment of a referential integrity constraint through the use of DDL prompts DBMS to offer discrete alternatives for addressing a violation that may arise from either a Delete or an Update operation (Pannirselvam et al., 1999).

3.4.4. The Transaction Concept

In the context of a relational database, it is common for a database application program to carry out one or multiple transactions Transaction refers to a program in execution that involves certain database operations, including reading from database or performing insertions, deletions, or updates to database. Upon transaction completion, it's imperative that database is left in valid & consistent state, which adheres to each of constraints outlined in database schema. Transaction can encompass multiple retrieval and update operations. The retrieval and update operations will collectively constitute a single, indivisible transactional unit performed on the database. In banking transactions, initiating a withdrawal from a user's account typically involves accessing the account record, verifying the availability of adequate funds, and subsequently modifying the record to reflect the withdrawal amount.

Online transaction processing (OLTP) systems used by many commercial applications that use relational databases are capable of processing transactions at rates surpassing several hundred per second (Chu & Ieong, 1993).

3.5. BENEFITS OF RELATIONAL DATABASE CONSTRAINTS

3.5.1. Ensuring Data Accuracy and Consistency

The principal advantage of incorporating relational database constraints lies in their capacity to guarantee precision and uniformity of data. Constraints aid in preserving the accuracy and consistency of the information contained within a database by imposing predetermined regulations and limitations. The imposition of constraints can play a significant role in enhancing the precision and uniformity of data:

1. **Uniqueness:** Limitations such as primary keys and unique keys guarantee the uniqueness of every entry within a given table. In the context of database management, a primary key refers to a specific column or a set of columns that serves the purpose of uniquely identifying each individual row within a given table. The primary key value is enforced as a rule to prevent the insertion of duplicate data, ensuring that no two rows possess the same value. Similarly, unique keys ensure that there are no duplicate values within a designated column or group of columns. Constraints prevent data redundancy and guarantee that each piece of information is accurately and uniquely represented within the database by preserving its uniqueness (Hill et al., 2013).
2. **Referential Integrity:** The constraint of referential integrity serves to institute and uphold connections among tables using foreign keys. A foreign key is a column or a set of columns in one table that refers to the primary key in another table. Constraints guarantee the coherence of data by imposing referential integrity, which mandates that the values contained in foreign key column of a given table correspond to values of the primary key in the table being referenced. By implementing this measure, orphaned records are avoided, and the coherence of data is maintained throughout interconnected tables. The referential integrity constraint guarantees a valid association between the orders

and customers' tables. Specifically, it verifies that the customer ID in the orders table corresponds to a valid customer ID in the customer's table (Barak et al., 2007).

3. Data Validation constraints, such as domain and check constraints, impose regulations on the permissible range and structure of data contained within columns. A domain constraint is a specification that defines a column's acceptable data types, lengths, formats, and other attributes. The function of data validation is to ensure that the information contained within a given column conforms to the predetermined criteria. An instance of a domain constraint can be observed where a particular column that denotes a date is restricted to only accept dates that conform to a predetermined range and format (Khalfi et al., 2017). Check constraints enable the implementation of intricate data validation measures that go beyond the scope of ensuring uniqueness and nullability. Organizations can define custom conditions to ensure data is inserted into the database only after meeting specific requirements. Exemplar employment of a check constraint is to guarantee that a specific column denoting an individual's age solely permits affirmative numerical values. Constraints validate data against predetermined criteria, thereby preventing the insertion of invalid or inconsistent data. This, in turn, helps to uphold the accuracy and dependability of the database (Mohanty et al., 2007).

3.5.2. Preventing Invalid and Inconsistent Data

Relational database constraints serve as protective measures to prevent the insertion of data that is invalid or inconsistent. Constraints are implemented to ensure that the database only accepts data that adheres to predetermined rules and limitations. This prevents the inclusion of data that fails to meet the specified criteria. Several methods exist by which constraints aid in the prevention of invalid and inconsistent data:

1. **Nullability Constraints:** The nullability constraints of a column dictate its ability to accept null values. The concept of null value pertains to the absence of data or an unknown value. Constraints guarantee the presence of crucial data by indicating whether a column permits null values or not. Implementing measures to prevent the inclusion of incomplete or unidentified data is crucial

in ensuring the accuracy of analyzes and the validity of outcomes. If a column denoting a customer's phone number disallows null values, the database shall implement the constraint and disapprove any endeavor to insert a record lacking a phone number (Kupzyk & Cohen, 2015).

2. **Key Constraints:** Primary and unique keys are essential constraints that restrict the insertion of repetitive values in particular columns or groups of columns. The primary key serves as a distinct identifier for every record in a table, whereas a unique key guarantees uniqueness within a designated column or group of columns.

   The database system guarantees the absence of duplicate primary key or unique key values across multiple rows by implementing key constraints. Eliminating data redundancy mitigates the possibility of duplicity and promotes the preservation of data consistency and precision. Suppose a table possesses a unique key constraint on the email column. In that case, the database will decline any endeavor to insert a record containing an email address that already exists within the table (Soares & Borba, 2002).

3. **Check Constraints:** Check constraints silently but powerfully preventing data discrepancies. Organizations can establish custom conditions to authenticate data prior to its insertion into database. An example of the utilization of check constraint is to guarantee that a particular column, which denotes a discount percentage, solely accepts values ranging from 0 to 100. The database system effectively averts the inclusion of data that fails to meet the prescribed criteria by implementing check constraints. This, in turn, precludes the existence of erroneous or incongruous data (Kent, 1983).

3.5.3. Supporting Data Integrity and Reliability

Implementing relational database constraints plays a crucial role in ensuring integrity & dependability of the data stored within database. Process of ensuring that the data conforms to predetermined rules and requirements serves to mitigate the possibility of errors and inconsistencies. The implementation of constraints can provide significant support for maintaining data integrity and reliability:

Remember

Relational database constraints play a crucial role in supporting data integrity and .reliability, ensuring the accuracy, consistency, and trustworthiness of stored data

1. **Data Consistency:** The implementation of constraints, such as primary keys, foreign keys, and referential integrity constraints, serves to guarantee the precise maintenance of data relationships. In order to avoid potential inconsistencies that may arise from incomplete or mismatched data references, measures are taken to prevent such occurrences. If foreign key is linked to primary key in separate table, the referential integrity constraint guarantees the presence of the referenced key within the referenced table. Data consistency is maintained throughout the database by preventing the creation of orphaned records or invalid references (Park et al., 2013).
2. **Error Prevention:** The DBMS ensures data integrity by imposing constraints during data insertion, update, and deletion operations, thereby preventing errors. Constraints serve as a mechanism for regulating data input into a database, permitting solely accurate and coherent information to be stored. Suppose a record is attempted to be inserted which violates a uniqueness constraint. In that case, the DBMS will decline the operation and generate an error, thereby obstructing the insertion of erroneous data (Almassabi et al., 2018).
3. **Data Quality and Reliability**: Implementing relational database constraints has been shown to enhance the overall quality and dependability of data. Constraints play a crucial role in maintaining accuracy, reliability, & suitability of the data stored in a database for critical business operations, reporting, and decision-making processes. This is achieved by preventing the insertion of invalid or inconsistent data. Organizations can rely on the data stored within the database, assuring that it conforms to predetermined regulations and limitations, thereby augmenting the credibility and practicality of the data (Azzini et al., 2019).

Activity 4.1

Design a relational database for a university's student information system. Outline the key considerations you would take into account when defining the relational model, including the characteristics of relations and the use of key constraints. Discuss how these constraints contribute to data integrity and reliability in managing .student data

3.6. SUMMARY

The present chapter expounded upon modeling concepts, data structures, & constraints that relational data model furnishes. The initial step involved the introduction of fundamental notions such as domains, tuples & attributes. Subsequently, relation schema was established as an inventory of attributes which delineate configuration of relation. A relation, also known as a relation state, can be defined as a collection of tuples that adhere to a specific schema.

Several distinguishing features set relations apart from conventional tables or files. One of the fundamental properties of a relation is its insensitivity to tuples ordering. Second aspect pertains to arrangement of attributes within relation schema & corresponding arrangement of values within tuple. An alternative conceptualization of the term "relation" was introduced, which doesn't require the existence of 2 distinct orderings. However, for the sake of convenience, we opted to employ the initial definition that mandates the ordering of attributes and tuple values. Subsequently, the topic of values within tuples was deliberated upon, wherein the concept of NULL values was introduced to signify the absence or uncertainty of information. It was emphasized that minimizing the usage of NULL values is advisable.

The categorization of database constraints includes three distinct types: inherent model-based constraints, explicit schema-based constraints, & application-based constraints, which're also referred to as semantic constraints or business rules. Subsequently, we engaged in a discourse regarding the limitations imposed by the schema.

In regards to the relational model, it is necessary to begin with domain constraints and subsequently address key constraints, which encompass the notions of superkey, candidate key, & primary key, as well as NOT NULL constraint applied to attributes. Relational databases & their corresponding schemas were established. One of the relational constraints that can be considered is entity integrity constraint. This constraint disallows primary key attributes from having a NULL value. Referential integrity constraint is a mechanism utilized to maintain references consistency among tuples that come from different relations, as explained in text.

Relational model encompasses three primary modification operations, namely Insert, Delete, & Update. Each individual operation has the potential to contravene specific categories of restrictions. It is imperative to verify that no constraints have been breached upon execution of an operation by examining the database state subsequent to the operation. The advent of the transaction concept was significant for relational database management systems because it made it possible to combine several database operations into one atomic database action.

REVIEW QUESTIONS

1. Define following terms as they apply to the relational model of data: domain, attribute, n-tuple, relation schema, relation state, degree of a relation, relational database schema, and relational database state.
2. Why are tuples in relation not ordered?
3. Why are duplicate tuples not allowed in a relation?
4. What is the difference between superkey & key?
5. What is the rationale behind selecting a specific candidate key as the primary key for a relation?
6. Outline the distinctive features of relations that set them apart from ordinary tables and files.
7. Discuss different reasons which lead to NULL values in relations.
8. Explain concepts of entity integrity and referential integrity constraints, and discuss their significance and importance in a database system.
9. Define foreign key. What is this concept used for?
10. Define transaction, and how does it distinguish itself from an update operation?

REFERENCES

1. Adya, A., Bolosky, W. J., Castro, M., Cermak, G., Chaiken, R., Douceur, J. R., & Wattenhofer, R. P., (2002). FARSITE: Federated, available, and reliable storage for an incompletely trusted environment. *ACM SIGOPS Operating Systems Review*, *36*(SI), 1–14.
2. Ai, T., & Zhang, Q., (2019). The application and research of cloud platform in practice teaching integrated service. In: *Journal of Physics: Conference Series* (Vol. 1213, pp. 042035). IOP Publishing.
3. Aleksić, S., Ristić, S., & Luković, I., (2011). *An Approach to Generating Server Implementation of the Inverse Referential Integrity Constraints* (Vol. 2, pp. 85, 86). Proceedings; AL-Zaytoonah University of Jordan, Amman, Jordan.
4. Almassabi, A., Bawazeer, O., & Adam, S., (2018). Top new SQL databases and features classification. *International Journal of Database Management Systems*, *10*(2), 11–31.
5. Arqués, D., & Michel, C., (1993). A model of gene evolution based on recognizable languages and on insertion and deletion operations. *International Journal of Modeling and Simulation*, *13*(3), 110–113.
6. Astrova, I., (2004). Reverse engineering of relational databases to ontologies. In: *The Semantic Web: Research and Applications: First European Semantic Web Symposium, ESWS 2004 Heraklion, Crete, Greece, May 10–12, 2004; Proceedings 1* (Vol. 2, pp. 327–341). Springer Berlin Heidelberg.
7. Atzeni, P., & Morfuni, N. M., (1986). Functional dependencies and constraints on null values in database relations. *Information and Control*, *70*(1), 1–31.
8. Azzini, A., Ceravolo, P., & Colella, M., (2019). Performances of OLAP operations in graph and relational databases. In: *Knowledge Management in Organizations: 14th International Conference, KMO 2019, Zamora, Spain, July 15–18, 2019, Proceedings 14* (Vol. 2, pp. 282–293). Springer International Publishing.
9. Barak, B., Chaudhuri, K., Dwork, C., Kale, S., McSherry, F., & Talwar, K., (2007). Privacy, accuracy, and consistency too: A holistic solution to contingency table release. In: *Proceedings of the Twenty-Sixth ACM SIGMOD-SIGACT-SIGART Symposium on Principles of Database Systems* (pp. 273–282).

10. Baumann, A., Heiser, G., Appavoo, J., Da Silva, D., Krieger, O., Wisniewski, R. W., & Kerr, J., (2005). Providing dynamic update in an operating system. In: *USENIX Annual Technical Conference, General Track* (Vol. 2, pp. 279–291).
11. Ben-Zvi, J., (1982). *The Time Relational Model* (Vol. 85, No. 5, p. 741). University of California, Los Angeles.
12. Bhattacharya, S., & Cortesi, A., (2009). A distortion free watermark framework for relational databases. In: *ICSOFT (2)* (Vol. 1, pp. 229–234).
13. Biskup, J., (1984). Extending the relational algebra for relations with maybe tuples and existential and universal null values. *Fundamenta Informaticae*, *7*(1), 129–150.
14. Boydens, I., Pirotte, A., & Zimányi, E., (1998). Managing constraint violations in administrative information systems. In: *Data Mining and Reverse Engineering: Searching for Semantics. IFIP TC2 WG2. 6 IFIP Seventh Conference on Database Semantics (DS-7) 7–10 October 1997, Leysin, Switzerland* (Vol. 1, pp. 353–377). Springer US.
15. Calero, C., Piattini, M., & Genero, M., (2001). Empirical validation of referential integrity metrics. *Information and Software Technology*, *43*(15), 949–957.
16. Candan, K. S., Grant, J., & Subrahmanian, V. S., (1997). A unified treatment of null values using constraints. *Information Sciences*, *98*(1–4), 99–156.
17. Capello, R., (2002). Spatial and sectoral characteristics of relational capital in innovation activity. *European Planning Studies*, *10*(2), 177–200.
18. Cerbah, F., (2008). Learning highly structured semantic repositories from relational databases: The RDBToOnto tool. *Lecture Notes in Computer Science*, *5021*, 777–781.
19. Chen, P. P. S., (1976). The entity-relationship model—Toward a unified view of data. *ACM Transactions on Database Systems (TODS)*, *1*(1), 9–36.
20. Chiang, R. H., Barron, T. M., & Storey, V. C., (1994). Reverse engineering of relational databases: Extraction of an EER model from a relational database. *Data & Knowledge Engineering*, *12*(2), 107–142.
21. Chu, W. W., & Ieong, I. T., (1993). A transaction-based approach to vertical partitioning for relational database systems. *IEEE Transactions on Software Engineering*, *19*(8), 804–812.

22. Clifford, J., & Croker, A., (1987). The historical relational data model (HRDM) and algebra based on lifespans. In: *1987 IEEE Third International Conference on Data Engineering* (Vol. 2, pp. 528–537). IEEE.
23. Codd, E. F., (1971). A data base sublanguage founded on the relational calculus. In: *Proceedings of the 1971 ACM SIGFIDET (now SIGMOD) Workshop on Data Description, Access, and Control* (Vol. 1, pp. 35–68).
24. Codd, E. F., (1972). Further normalization of the data base relational model. *Data Base Systems*, *6*, 33–64.
25. Codd, E. F., (1979). Extending the database relational model to capture more meaning. *ACM Transactions on Database Systems (TODS)*, *4*(4), 397–434.
26. Cunha, J., Saraiva, J., & Visser, J., (2009). From spreadsheets to relational databases and back. In: *Proceedings of the 2009 ACM SIGPLAN Workshop on Partial Evaluation and Program Manipulation* (Vol. 4, pp. 179–188).
27. Date, C. J., (1982). A formal definition of the relational model. *ACM SIGMOD Record*, *13*(1), 18–29.
28. Dayal, U., & Bernstein, P. A., (1982). On the correct translation of update operations on relational views. *ACM Transactions on Database Systems (TODS)*, *7*(3), 381–416.
29. Delobel, C., (1978). Normalization and hierarchical dependencies in the relational data model. *ACM Transactions on Database Systems (TODS)*, *3*(3), 201–222.
30. Demuth, B., & Hussmann, H., (1999). Using UML/OCL constraints for relational database design. In: *«UML» '99—The Unified Modeling Language: Beyond the Standard Second International Conference Fort Collins, CO, USA, October 28–30, 1999 Proceedings 2* (Vol. 1, pp. 598–613) Springer Berlin Heidelberg.
31. Dey, D., & Sarkar, S., (1996). A probabilistic relational model and algebra. *ACM Transactions on Database Systems (TODS)*, *21*(3), 339–369.
32. Di Nitto, E., & Tanca, L., (1996). Dealing with deviations in DBMSs: An approach to revise consistency constraints. In: *FMLDO* (Vol. 1, pp. 11–24).
33. Eckert, F., Fort, T. C., Schott, P. K., & Yang, N. J., (2020). *Imputing Missing Values in the US Census Bureau's County Business Patterns* (Vol. 3, pp. 741–743). National Bureau of Economic Research.

34. Elmasri, R., Weeldreyer, J., & Hevner, A., (1985). The category concept: An extension to the entity-relationship model. *Data & Knowledge Engineering, 1*(1), 75–116.

35. Fahrner, C., & Vossen, G., (1995). Transforming relational database schemas into object-oriented schemas according to ODMG-93. In: *Deductive and Object-Oriented Databases: Fourth International Conference, DOOD'95 Singapore, December 4–7, 1995 Proceedings 4* (Vol. 1, pp. 429–446). Springer Berlin Heidelberg.

36. Fonkam, M. M., & Gray, W. A., (1992). An approach to eliciting the semantics of relational databases. In: *Advanced Information Systems Engineering: 4th International Conference CAiSE'92 Manchester, UK, May 12–15, 1992 Proceedings 4* (Vol. 1, pp. 463–480). Springer Berlin Heidelberg.

37. Gadia, S. K., (1988). A homogeneous relational model and query languages for temporal databases. *ACM Transactions on Database Systems (TODS), 13*(4), 418–448.

38. Galan, J., (2020). Towards a relational model for emerging urban nature concepts: A practical application and an external assessment in landscape planning education. *Sustainability, 12*(6), 2465.

39. Gehani, N. H., & Jagadish, H. V., (1991). Ode as an active database: Constraints and triggers. In: *VLDB* (Vol. 91, pp. 327–336).

40. Getoor, L., & Sahami, M., (1999). Using probabilistic relational models for collaborative filtering. In: *Workshop on Web Usage Analysis and User Profiling (WEBKDD'99)* (Vol. 2, pp. 1–6).

41. Goldstein, B. S., (1981). Constraints on null values in relational databases. In: *Proceedings of the Seventh International Conference on Very Large Data Bases* (Vol. 1, 7, pp. 101–110).

42. Halpin, T. A., & Ritson, P. R., (1992). Fact-oriented modelling and null values. *Research and Practical Issues in Databases, 321*, 1122–1123.

43. Halpin, T., & Morgan, T., (2010). *Information Modeling and Relational Databases* (Vol. 6, No. 6, p. 990–1000). Morgan Kaufmann.

44. He, J. S. K. T. G. & Naughton, C., (2008). Relational databases for querying XML documents: Limitations and opportunities. *Proceedings of VLDB*, 302–314.

45. Hill, C., Gordon, I. E., Rothman, L. S., & Tennyson, J., (2013). A new relational database structure and online interface for the HITRAN

database. *Journal of Quantitative Spectroscopy and Radiative Transfer*, *130*, 51–61.

46. Hui, K. Y., & Gray, P. M., (2000). Developing finite domain constraints-a data model approach. *Lecture Notes in Computer Science*, *78*, 448–462.
47. Jajodia, S., & Sandhu, R., (1991). Toward a multilevel secure relational data model. *ACM SIGMOD Record*, *20*(2), 50–59.
48. Kappel, G., Kapsammer, E., & Retschitzegger, W., (2001). XML and relational database systems-A comparison of concepts. In: *International Conference on Internet Computing (1)* (Vol. 3, pp. 199–205).
49. Kent, W., (1983). A simple guide to five normal forms in relational database theory. *Communications of the ACM*, *26*(2), 120–125.
50. Khalfi, B., De Runz, C., Faiz, S., & Akdag, H., (2017). A new methodology for storing consistent fuzzy geospatial data in big data environment. *IEEE Transactions on Big Data*, *7*(2), 468–482.
51. Kupzyk, K. A., & Cohen, M. Z., (2015). Data validation and other strategies for data entry. *Western Journal of Nursing Research*, *37*(4), 546–556.
52. Lerat, N., & Lipski, Jr. W., (1986). Nonapplicable nulls. *Theoretical Computer Science*, *46*, 67–82.
53. Levene, M., & Loizou, G., (2001). A generalization of entity and referential integrity in relational databases. *RAIRO-Theoretical Informatics and Applications*, *35*(2), 113–127.
54. Levene, M., & Loizou, G., (2012). *A Guided Tour of Relational Databases and Beyond* (Vol. 2, pp. 253–255). Springer Science & Business Media.
55. Li, M., Du, X. Y., & Wang, S., (2005). Learning ontology from relational database. In: *2005 International Conference on Machine Learning and Cybernetics* (Vol. 6, pp. 3410–3415). IEEE.
56. Lin, T. Y., (1988). Neighborhood systems and approximation in relational databases and knowledge bases. In: *Proceedings of the 4th International Symposium on Methodologies of Intelligent Systems* (Vol. 2, pp. 75–86). California State University Northridge CA.
57. Liu, Z., Jifeng, H., Li, X., & Chen, Y., (2003). A relational model for formal object-oriented requirement analysis in UML. In: *Formal Methods and Software Engineering: 5th International Conference on Formal Engineering Methods, ICFEM 2003, Singapore, November 5–7, 2003. Proceedings 5* (Vol. 3, pp. 641–664). Springer Berlin Heidelberg.

58. Luković, I., Mogin, P., Pavićević, J., & Ristić, S., (2007). An approach to developing complex database schemas using form types. *Software: Practice and Experience, 37*(15), 1621–1656.
59. Maatuk, A., Ali, A., & Rossiter, N., (2008). Relational database migration: A perspective. In: *Database and Expert Systems Applications: 19th International Conference, DEXA 2008, Turin, Italy, September 1–5, 2008; Proceedings 19* (Vol. 2, pp. 676–683). Springer Berlin Heidelberg.
60. Malinowski, E., & Zimányi, E., (2008). A conceptual model for temporal data warehouses and its transformation to the ER and the object-relational models. *Data & Knowledge Engineering, 64*(1), 101–133.
61. Marcos, E., Vela, B., & Cavero, J. M., (2003). A methodological approach for object-relational database design using UML. *Software and Systems Modeling, 2*, 59–72.
62. Markowitz, V. M., & Makowsky, J. A., (1990). Identifying extended entity-relationship object structures in relational schemas. *IEEE Transactions on Software Engineering, 16*(8), 777–790.
63. McMinn, P., Wright, C. J., McCurdy, C. J., & Kapfhammer, G. M., (2017). Automatic detection and removal of ineffective mutants for the mutation analysis of relational database schemas. *IEEE Transactions on Software Engineering, 45*(5), 427–463.
64. Meier, A., Schindler, G., & Werro, N., (2008). Fuzzy classification on relational databases. In: *Handbook of Research on Fuzzy Information Processing in Databases* (Vol. 1, pp. 586–614). IGI Global.
65. Meurice, L., Ruiz, F. J. B., Weber, J. H., & Cleve, A., (2014). Establishing referential integrity in legacy information systems-reality bites! In: *2014 IEEE International Conference on Software Maintenance and Evolution* (Vol. 1, pp. 461–465). IEEE.
66. Mohanty, S. K., Piccoli, A. L., Devine, L. J., Patel, A. A., William, G. C., Winters, S. B., & Parwani, A. V., (2007). Synoptic tool for reporting of hematological and lymphoid neoplasms based on World Health organization classification and college of American pathologists checklist. *BMC Cancer, 7*, 1–9.
67. Ng, W., & Levene, M., (1997). An extension of SQL to support ordered domains in relational databases. In: *Proceedings of the 1997 International Database Engineering and Applications Symposium (Cat. No. 97TB100166)* (Vol. 1, pp. 358–367). IEEE.

68. Nyulas, C., O'connor, M., & Tu, S., (2007). DataMaster–a plug-in for importing schemas and data from relational databases into Protege. In: *10th International Protégé Conference* (Vol. 2, pp. 15–18).

69. Pannirselvam, G. P., Ferguson, L. A., Ash, R. C., & Siferd, S. P., (1999). Operations management research: An update for the 1990s. *Journal of Operations Management*, *18*(1), 95–112.

70. Paredaens, J., De Bra, P., Gyssens, M., & Van, G. D., (2012). *The Structure of the Relational Database Model* (Vol. 17, pp. 981, 982). Springer Science & Business Media.

71. Park, S., Kelly, T., & Shen, K., (2013). Failure-atomic msync() a simple and efficient mechanism for preserving the integrity of durable data. In: *Proceedings of the 8th ACM European Conference on Computer Systems* (Vol. 2, pp. 225–238).

72. Pieris, D., (2013). *Modifying the Entity Relationship Modeling Notation: Towards High Quality Relational Databases from Better Notated ER Models* (Vol. 445, pp. 852–855). arXiv preprint arXiv.

73. Raghunathan, T. E., Lepkowski, J. M., Van, H. J., & Solenberger, P., (2001). A multivariate technique for multiply imputing missing values using a sequence of regression models. *Survey Methodology*, *27*(1), 85–96.

74. Reinert, J., (1995). Ambiguity for referential integrity is undecidable. In: *ESPRIT WG CONTESSA Workshop on Constraint Databases and Applications* (Vol. 1, pp. 132–147). Berlin, Heidelberg: Springer Berlin Heidelberg.

75. Scannapieco, M., & Batini, C., (2004). Completeness in the relational model: A comprehensive framework. In: *ICIQ* (Vol. 1, pp. 333–345).

76. Schek, H. J., & Scholl, M. H., (1986). The relational model with relation-valued attributes. *Information Systems*, *11*(2), 137–147.

77. Selinger, P. G., Astrahan, M. M., Chamberlin, D. D., Lorie, R. A., & Price, T. G., (1979). Access path selection in a relational database management system. In: *Proceedings of the 1979 ACM SIGMOD International Conference on Management of Data* (Vol. 1, pp. 23–34).

78. Shenoi, S., & Melton, A., (1989). Proximity relations in the fuzzy relational database model. *Fuzzy Sets and Systems*, *31*(3), 285–296.

79. Soares, S., & Borba, P., (2002). Concurrency control with Java and relational databases. In: *Proceedings 26th Annual International Computer Software and Applications* (Vol. 2, pp. 843–849). IEEE.

80. Sumathi, S., & Esakkirajan, S., (2007). *Fundamentals of Relational Database Management Systems.* (Vol. 47, pp. 789, 790). Springer.
81. Tansel, A. U., (1986). Adding time dimension to relational model and extending relational algebra. *Information Systems, 11*(4), 343–355.
82. Tatarinov, I., Viglas, S. D., Beyer, K., Shanmugasundaram, J., Shekita, E., & Zhang, C., (2002). Storing and querying ordered XML using a relational database system. In: *Proceedings of the 2002 ACM SIGMOD International Conference on Management of Data* (Vol. 1, pp. 204–215).
83. Telnarova, Z., (2010). Relational database as a source of ontology creation. In: *Proceedings of the International Multiconference on Computer Science and Information Technology* (Vol. 1, pp. 135–139). IEEE.
84. Thapa, R. B., & Giese, M., (2022). *Mapping Relational Database Constraints to SHACL (Extended Version), 1, 4-8* .
85. Vidaković, J., Luković, I., & Kordić, S., (2016). Specification and validation of the referential integrity constraint in XML databases. In: *Proceedings of the 6th International Conference on Information Society and Technology* (Vol. 3, pp. 197–202).
86. Wand, Y., Storey, V. C., & Weber, R., (1999). An ontological analysis of the relationship construct in conceptual modeling. *ACM Transactions on Database Systems (TODS), 24*(4), 494–528.
87. Wu, J., Xiong, R., & Chiclana, F., (2016). Uninorm trust propagation and aggregation methods for group decision making in social network with four tuple information. *Knowledge-Based Systems, 96*, 29–39.
88. Yan, L. L., & Ling, T. W., (1993). Translating relational schema with constraints into OODB schema. In: *Interoperable Database Systems (Ds-5)* (Vol. 2, pp. 69–85). North-Holland.
89. Yue, K. B., (2022). Assuring referential integrity in blockchain applications. In: *2022 Fourth International Conference on Blockchain Computing and Applications (BCCA)* (Vol. 1, pp. 47–54). IEEE.
90. Zaniolo, C., (1982). Database relations with null values. In: *Proceedings of the 1st ACM SIGACT-SIGMOD Symposium on Principles of Database Systems* (Vol. 1, pp. 27–33).
91. Zvieli, A., & Chen, P. P., (1986). Entity—Relationship modeling and fuzzy databases. In: *1986 IEEE Second International Conference on Data Engineering* (Vol. 1, pp. 320–327). IEEE.

CHAPTER

4

Extract, Transform, Load (ETL) Processing with SQL

CONTENT

4.1. INTRODUCTION

The ETL process and its associated activities involve the extraction of vital information from other sources, the transformation of data into a polished, consistent, and useful format, and the incorporation of said data through the data warehouse. This procedure holds considerable significance in guaranteeing the effectiveness of an individual's Business Intelligence (BI) solutions. Proficiency in data management is a crucial determinant of the operational efficiency and practicality of a data center. A disorganized as well as unfit data warehouse can be distinguished from a well-managed one based on this factor. The execution of Extract, Transform, and Load (ETL) operations is a significant and intricate phase in the development of a BI system (Galici, Ordile, Marchesi, Pinna, & Tonelli, 2020) (Figure 4.1).

Figure 4.1. Illustration of ETL components.

Source: Informatica, Creative Commons License.

The present chapter expounds upon the procedure of carrying out the ETL procedures that are imperative for the efficacious deployment of one's BI resolution. The main aim of the forthcoming sections is to demonstrate a technique that utilizes a combination of SQL programming and SQL Server Integration Service (SSIS) to construct an effective ETL project, which will serve as a crucial element of your BI system (Wang & Ye, 2010). The present chapter discusses the commonly used SQL programming methodologies that are utilized for identifying and resolving issues that are associated with the process of ETL. It is possible to propose a method for transforming the codes used in SQL methods into views that are saved procedures, which can then be employed in the process of ETL comprehension (Awiti, Vaisman, & Zimányi, 2020) (Figure 4.2).

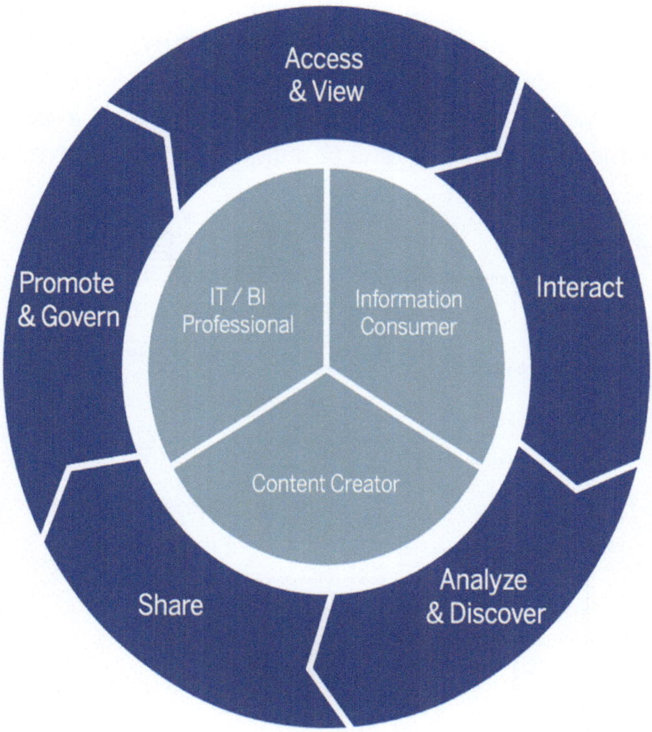

Figure 4.2. Schematic of components of BI.

Source: Tableau, Creative Commons License.

4.2. PERFORMING THE ETL PROGRAMMING

The process of developing an ETL system utilizing a blend of SQL Server as well as SSIS can be depicted by the sequential procedures illustrated in Figure 4.3. It should be noted that the process involves making a decision regarding whether to populate a table entirely with new data, to load it incrementally (as elaborated in the subsequent section), or to update any modifications from the primary source (Gour, Sarangdevot, Tanwar, & Sharma, 2010).

The aforementioned procedure involves identifying the data to be extracted and scrutinizing its contents for accuracy, consistency, and comprehensiveness. Upon confirming that the information at hand satisfies the requirements, it is probable that some degree of manipulation may still be necessary to conform to the location tables. The act of manipulating data may manifest as the alteration of column names or the conversion of source data types into their respective target data types. After completing all necessary arrangements, the data may be loaded into the designated data center tables, as well as the aforementioned process may be repeated for each subsequent table that requires a data population (Kakish & Kraft, 2012).

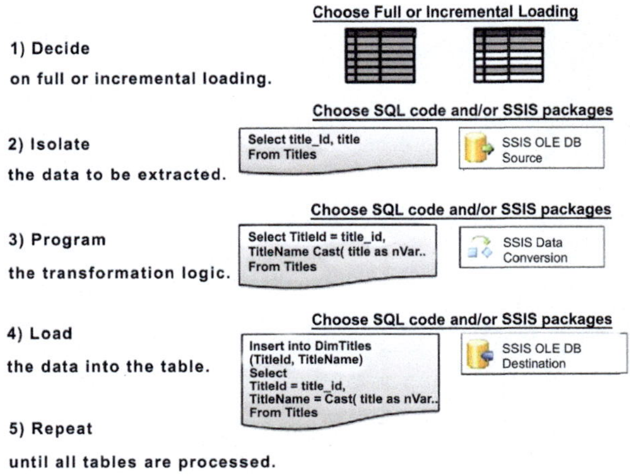

Figure 4.3. Illustration of extract, transform, and load process utilizing SQL Server as well as SQL server integration services.

Source: Science direct, Creative Commons License.

The majority of these procedures can be executed by using SQL queries or just SSIS tasks, both of which are analyzed in this publication. A more comprehensive comprehension of the SSIS tasks' function can be achieved by initially scrutinizing the SQL statements which they embody. Hence, it is imperative to analyze the SQL programming commands required to accomplish these procedures (Diouf, Boly, & Ndiaye, 2017).

4.3. CHOOSING BETWEEN FULL LOADING AND LOADING IN STAGES

The tables within the data storeroom have the option to undergo either a complete truncation and repopulation process or an incremental loading process. The technique of emptying the tables as well as subsequently replenishing them in their entirety is commonly referred to by the term a flush and fill method. This methodology represents a straightforward approach for executing an ETL procedure, albeit its efficacy diminishes when applied to sizable datasets. The flush and fill method is effective in scenarios where dimension tables are of relatively small size, typically comprising a limited number of thousand rows (Vassiliadis, Simitsis, & Baikousi, 2009). Tables of significant size, like the fact table, can contain a vast number of rows, potentially numbering in the millions. Hence, the duration required to entirely evacuate the table and subsequently replenish it with updated information may be unduly prolonged. In such instances, opting to solely input modified data from the primary source is a significantly more effective decision (Svilar, Castellano, Jukic, & Casamichana, 2018).

The flush and fill method involves the clearance of tables within a SQL Server database's data warehouse through the utilization of one of two things: the DELETE or just the TRUNCATE command. Upon executing the delete command, individual rows are systematically removed from the table. Thus, in the event that a table contains 1,000 rows, the deletion operation will be executed 1,000 times. The process occurs expeditiously, albeit necessitating a duration longer than that of a mere truncation. The TRUNCATE command is utilized in SQL Server for removing data pages that are responsible for storing data internally (Simitsis, 2004). Subsequently, these imperceptible data pages can be repurposed for alternative entities within a SQL Server Databases. The expeditious method of emptying a SQL table is through truncation, however, the presence of foreign key restrictions on said table prohibits its truncation. An instance presented in Listing 4–1 illustrates the syntax of the DELETE command in SQL (Theodorou, Abelló, Lehner, & Thiele, 2016).

Listing 4–1. The process of removing info from tables within a data warehouse can be accomplished through the use of the delete command.

```
Delete From dbo.FactSales
Delete From dbo.FactTitlesAuthors
Delete From dbo.DimTitles
Delete From dbo.DimPublishers
Delete From dbo.DimStores
Delete From dbo.DimAuthors
```

In the event that the truncation assertion is utilized in this context, it is imperative that the source code incorporates assertions that eliminate the foreign key connections prior to truncation. The TRUNCATE instruct is exemplified in Listing 4–2, showcasing the appearance of SQL code.

Listing 4–2. Getting rid of all the table data while starting over with the identity values.

/****** Drop Foreign Key s ******/

Alter Table [dbo].[DimTitles] Drop Constraint [FK_DimTitles_DimPublishers]

Alter Table [dbo].[FactTitlesAuthors] Drop Constraint [FK_FactTitlesAuthors_DimAuthors]

Alter Table [dbo].[FactTitlesAuthors] Drop Constraint [FK_FactTitlesAuthors_DimTitles]

Alter Table [dbo].[FactSales] Drop Constraint [FK_FactSales_DimStores] Alter Table [dbo].[FactSales] Drop Constraint [FK_FactSales_DimTitles] Go

/****** Clear all tables and reset their Identity Auto Number ******/

Truncate Table dbo.FactSales

Truncate Table dbo.FactTitlesAuthors

Truncate Table dbo.DimTitles

Truncate Table dbo.DimPublishers

Truncate Table dbo.DimStores

Truncate Table dbo.DimAuthors

Go

/****** Add Foreign Keys ******/

Alter Table [dbo].[DimTitles] With Check Add Constraint [FK_DimTitles_DimPublishers]

Foreign Key ([PublisherKey]) References [dbo].[DimPublishers] ([PublisherKey])

Alter Table [dbo].[FactTitlesAuthors] With Check Add Constraint [FK_FactTitlesAuthors_DimAuthors] Foreign Key ([AuthorKey]) References [dbo].[DimAuthors] ([AuthorKey])

Alter Table [dbo].[FactTitlesAuthors] With Check Add Constraint [FK_FactTitlesAuthors_DimTitles]

Foreign Key ([TitleKey]) References [dbo].[DimTitles] ([TitleKey])

Alter Table [dbo].[FactSales] With Check Add Constraint [FK_FactSales_DimStores] Foreign Key ([StoreKey]) References [dbo].[DimStores] ([Storekey])

Alter Table [dbo].[FactSales] With Check Add Constraint [FK_FactSales_DimTitles]

Foreign Key ([TitleKey]) References [dbo].[DimTitles] ([TitleKey]) Go

Truncation offers an advantage over deletion because it facilitates the automatic resetting of the numbering system to its initial value (usually 1) when a table utilizing an identifiable option to generate integer key values is involved. In contrast to resetting the number, deletion does not result in the resetting of the integer's value. Consequently, upon inserting a new row subsequent to deletion, the integer's value will resume from the point at which the previous inserts ceased. Typically, this is not a desirable outcome as the numerical sequence will deviate from commencing at 1, potentially leading to perplexity (Prabhu & Vēṅkaṭēcaṉ, 2007).

The code snippets presented in Listings 4–1 and 4–2 serve as illustrations of the implementation of the flush & fill procedure. However, in case of opting for an incremental load, it is advisable not to truncate the tables beforehand. Conversely, conduct a comparison of the values present in both the source along with destination tables. Subsequently, the process involves either appending rows to the target tables in instances where fresh rows are detected in the source, modifying rows in the target that have been altered within the source, or eliminating rows from the target that have been deleted within the source (SABTU et al., 2017).

The OLTP table of a customer within the given instance features a column named RowStatus, which serves as a flag indicator. This is illustrated in Listing 4–4. Upon the addition, update, or deletion of a row within an OLTP table, a flag is established to denote the corresponding operation. The examination of flags is conducted to ascertain the data that necessitates synchronization across the DimCustomers table from the Clients table (Prasath & Sreemathy, 2021). Subsequently, an INSERT, UPDATE, or just DELETE operation is executed based on the flag identified within the table.

Listing 4–4. Syncing Tables

```
Use TEMPDB

Go

-- Step #1. Make two demo tables

Create Table Customers

(CustomerId int

, CustomerName varchar(50)
```

```
, RowStatus Char(1) check(RowStatus in ('i,"u,"d')))
Go
Create Table DimCustomers
(CustomerId int
, CustomerName varchar(50))
Go
-- Step #2. Add some starting data
Insert into Customers (CustomerId, CustomerName, RowStatus)
Values(1, 'Bob Smith,' 'i')
Go
Insert into Customers (CustomerId, CustomerName, RowStatus)
Values(2, 'Sue Jones,' 'i')
Go
-- Step #4. Verify that the tables are not synchronized
Select * from Customers
Select * from DimCustomers
Go
-- Step #4 Synchronize the tables with this code
BEGIN TRANSACTION
Insert into DimCustomers
(CustomerId, CustomerName)
Select CustomerId, CustomerName
From Customers
Where RowStatus is NOT null
AND RowStatus='i'
-- Synchronize Updates
Update DimCustomers
Set DimCustomers.CustomerName = Customers.CustomerName
From DimCustomers
JOIN Customers
On DimCustomers.CustomerId = Customers.CustomerId
```

```
AND RowStatus = 'u'
-- Synchronize Deletes
Delete DimRows From DimCustomers as DimRows
JOIN Customers
On DimRows.CustomerId = Customers.CustomerId
AND RowStatus = 'd'
-- After we import data to the dim table -- we must reset the flags to null!
Update Customers Set RowStatus = null
COMMIT TRANSACTION
-- Step #5. Test that both tables now contain the same rows
Select * from Customers
Select * from DimCustomers
Go
-- Step #4. Test the Updates and Delete options
Update Customers
Set
CustomerName = 'Robert Smith'
, RowStatus = 'u'
Where CustomerId = 1
Go
Update Customers
Set
CustomerName = 'deleted'
, RowStatus = 'd'
Where CustomerId = 2
Go
-- Step #7. Verify that the tables are not synchronized
Select * from Customers
Select * from DimCustomers
Go
-- Step #8. Synchronize the tables with the same code as before
```

```
BEGIN TRANSACTION
Insert into DimCustomers
(CustomerId, CustomerName) Select CustomerId, CustomerName
From Customers
Where RowStatus is NOT null
AND RowStatus = 'i'
-- Synchronize Updates
Update DimCustomers
Set DimCustomers.CustomerName = Customers.CustomerName
From DimCustomers
JOIN Customers
On DimCustomers.CustomerId = Customers.CustomerId
AND RowStatus = 'u'
-- Synchronize Deletes
Delete DimRows From DimCustomers as DimRows
JOIN Customers
On DimRows.CustomerId = Customers.CustomerId
AND RowStatus = 'd'
-- After we import data to the dim table -- we must reset the flags to null!
Update Customers Set RowStatus = null
COMMIT TRANSACTION
-- Step #9. Test that both tables contain the same rows
Select * from Customers
Select * from DimCustomers
Go
-- Step #10. Setup an ETL process that will run the Synchronization code
```

As evident, formulating SQL code for achieving incremental loading may prove considerably intricate. Fortunately, this task is not required for the majority of tables. In instances where tables are insufficiently sized to derive advantages via the incremental method, it is advisable to maintain a simplistic ETL processing methodology and adhere to the flush & fill method. The book's emphasis is on using the flush & fill method for all tables within all 4 demo databases due to the minimal amount of info contained within them (Perron, Castro Fernandez, DeWitt, & Madden, 2020).

4.4. GETTING THE DATA THAT NEEDS TO BE RETRIEVED

It is imperative to analyze the requisite data to perform the ETL procedure. Initiating the process by retrieving all the available data via the table under consideration is a commendable approach. One can initiate the process by opening a query window, inputting a basic SELECT statement, as well as carrying out it to obtain the outcomes (Vyas & Vaishnav, 2017). The method is initiated by a statement, as exemplified in Listing 4–4.

Listing 4–4. Taking all of the information from the original table

```
Select * from [Pubs].[dbo].[Titles]
```

Setting Up Your Code

Frequently, the code employed for data isolation is subsequently utilized in the ETL (Extract, Transform, Load) procedure. As a consequence, it may be advisable to allocate sufficient time to ensure that your code exhibits a professional appearance. An effective approach to enhancing the professionalism of a person's code is to structure the ETL code within a proper format. The utilization of a star sign to represent all columns essentially is deemed acceptable over queries that are ad hoc (Rodzi, Othman, & Yusuf, 2015). However, a preferable approach is to specifically enumerate oneself columns, as demonstrated in Listing 4–5.

Listing 4–5. Outlining the Columns Clearly

```
Select
[title_id]
, [title]
, [type]
, [pub_id]
, [price]
, [advance]
, [royalty]
, [ytd_sales]
, [notes]
```

, [pubdate]

From [Pubs].[dbo].[Titles]

Square brackets are employed to enclose column as well as table names, as may be observed. Although optional, this standard is widely regarded as a best practice. An additional standard in academic writing involves the identification of a table by its complete or nearly complete name. The nomenclature convention for a table typically comprises four components, namely <ServerName>, <DatabaseName>, <SchemaName>, and <TableName>. The convention of utilizing solely the final three components of the completely qualified name seems widespread, whereas the server name component is infrequently employed (Bouman, Casters, & Van Dongen, 2010). By doing so, it implies the intention to retrieve data from a table located on a server that is not local. While this approach may offer benefits, it requires the configuration of linked servers within SQL Server, a practice that is infrequently implemented on an operational server. In the absence of a linked server, incorporating the server name into the complete table name produces an error. Consequently, the trinomial nomenclature is predominantly utilized (Karakasidis, Vassiliadis, & Pitoura, 2005).

4.5. TRANSFORMING LOGIC'S IDENTIFICATION

The ETL process typically involves a time-intensive phase of determining the necessary transformation and subsequently devising the corresponding programming logic. Examining an instance can aid in comprehending the components involved (Natarajan, 2011). Figure 4.4 displays two distinct tables, namely the Titles OLTP table for the database along with the DimTitles OLAP information warehouse table. The following analysis aims to compare and contrast the two tables in question.

1. The DimTitles table contains a reduced number of columns.
2. There exists a disparity in the column names utilized in the tables.
3. There exists a disparity in the data forms of certain columns.
4. A substitute key has been introduced to facilitate external key references.
5. The nullability of numerous columns is now altered.
6. Certain values have undergone a process of cleansing and enhancement to improve their readability (Seenivasan, 2023).

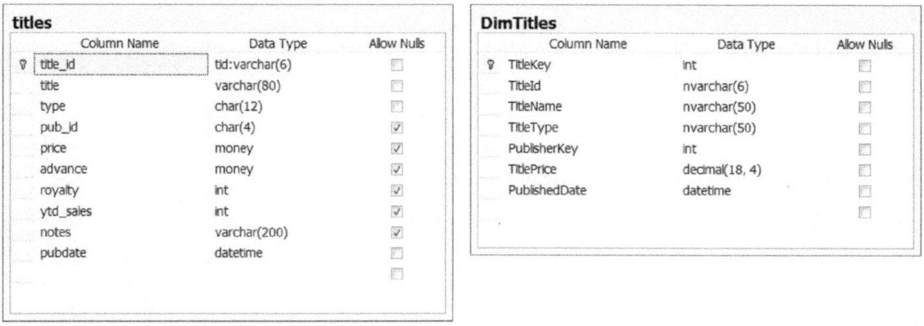

titles

Column Name	Data Type	Allow Nulls
title_id	tid:varchar(6)	☐
title	varchar(80)	☐
type	char(12)	☐
pub_id	char(4)	☑
price	money	☑
advance	money	☑
royalty	int	☑
ytd_sales	int	☑
notes	varchar(200)	☑
pubdate	datetime	☐

DimTitles

Column Name	Data Type	Allow Nulls
TitleKey	int	☐
TitleId	nvarchar(6)	☐
TitleName	nvarchar(50)	☐
TitleType	nvarchar(50)	☐
PublisherKey	int	☐
TitlePrice	decimal(18, 4)	☐
PublishedDate	datetime	☐

Figure 4.4. *Image of titles table and DimTitles comparison.*

Source: Geeks for Geeks, Creative Commons License.

The variances identified constitute a series of modifications that necessitate resolution during the transfer of data through the Titles table into the DimTitles table.

4.6. DEVELOPING YOUR TRANSFORMATION ALGORITHM

It is necessary to develop coding and programming frameworks to facilitate the transformation of any data which necessitates such action. The process of transformation may be executed through the utilization of SQL, software code (e.g., C#), or a hybrid approach that incorporates both.

The present code has been produced through the utilization of software tools like SSIS. Notwithstanding, code that develops automatically might not be properly optimized, thereby necessitating the need for manual optimization (Jain, Liao, & Willke, 2013).

In certain instances, it may be necessary to rectify the issue prior to utilizing it in a production environment. As one gains experience in ETL processing, it becomes increasingly evident that possessing a comprehensive comprehension of the code responsible for executing the transformations is crucial for proficiently generating SSIS packages.

The utilization of software utilities for code generation and proficiency in code optimization are two interdependent facets of ETL processing. The present discourse aims to examine prevalent programming methodologies that are facile to execute, yet yield substantial advantages for one's endeavors (Wojciechowski, 2011).

4.6.1. Data Reduction

It is improbable that the necessity to extract all columns via the source table will arise. Hence, the omission of undesired columns via the select section suffices. This elementary procedure constitutes the initial step in enhancing the efficiency of your (ETL method. The code presented in Listing 4–4 requests solely the columns that are necessary to populate the DimTitles table. Upon comparing the columns which are listed with the columns of the Titles table as depicted in Figure 4.4, it can be observed that there is a reduction of approximately 40% when compared to selecting nearly every column through a query like SELECT * FROM Titles (Bala, Boussaid, & Alimazighi, 2014).

Did You Know?

Data reduction techniques in the ETL process can significantly reduce storage .requirements, with compression ratios of up to 80% achieved in some cases

Listing 4–4. The process of exclusively choosing. The necessary information for the target table.

```
Select
[title_id]
, [title]
, [type]
, [pub_id]
, [price]
, [pubdate]
From [Pubs].[dbo].[Titles]
```

4.6.2. The Use of Column Aliases

It is advisable to utilize a column alias to make sure that the names of the source columns correspond with those of the target columns, thereby facilitating a change in the name of the former. This facilitates the identification of the relationship between one's sources and their destinations by others and also assists in the configuration of SSIS, which will be discussed in the subsequent chapter. SQL Server provides a pair of formats for creating column aliases: the initial format is [column name] AS [alias], along with the second format is [alias]=[column name] (Mutanu & Khan, 2021). The two of these methods achieve the same objective alongside are demonstrated in Listing 4–7.

Tip

Utilize column aliases in your SQL transformations to assign meaningful names to the output columns, improving the clarity and understandability of the transformed .data

Listing 4–7. Employing Various Techniques for Column Aliases

```
-- Older style column aliases: [column name] as [alias]
Select
[title_id] as[TitleId], [title] as [TitleName]
, [type] as [TitleType]
--, [pub_id] Will be replaced with a PublisherKey
, [price] as [TitlePrice]
, [pubdate] as [PublishedDate]
From [Pubs].[dbo].[Titles]
-- Newer style column aliases: [alias] = [column name]
Select
[TitleId] = [title_id]
, [TitleName] = [title]
, [TitleType] = [type]
--, [pub_id] Will be replaced with a PublisherKey
, [TitlePrice] = [price]
, [PublishedDate] = [pubdate]
From [Pubs].[dbo].[Titles]
```

4.6.3. Data Types Conversion

One of the transformations that may occur is the conversion of data types between the source as well as destination tables. The Microsoft SQL code presented in Listing 4–8 demonstrates the conversion of data types during the selection of data collected from its original table. This method offers a rapid and straightforward approach to executing such conversions (Trujillo & Luján-Mora, 2003).

Listing 4–8. Utilizing the Data Type Conversion Function

Select

[TitleId]=**Cast**([title_id] **as nvarchar(4))**

, [TitleName]=**Cast**([title] **as nvarchar(50))**

, [TitleType]=**Cast**([type] **as nvarchar(50))**

--, [pub_id] Will be replaced with a PublisherKey

, [TitlePrice]=**Cast**([price] **as decimal(18, 4))**

, [PublishedDate]=[pubdate] -- has the same data type in both tables

From [Pubs].[dbo].[Titles]

Those with prior experience in SQL might have observed that the aforementioned listing employs the CAST() act to alter the data kinds of pre-existing values in the original table. The cast operation serves as a rudimentary and long-standing function within the SQL Server database management system. A related function known as CONVERT() is capable of executing the aforementioned transformations. The CONVERT() act offers supplementary alternatives that do not exist within the CAST() purpose (El-Sappagh, Hendawi, & El Bastawissy, 2011). An instance of this phenomenon is the utilization by the CONVERT() functions to alter the structure of the date. The provided illustration, denoted as Figure 4.5, presents an exemplar.

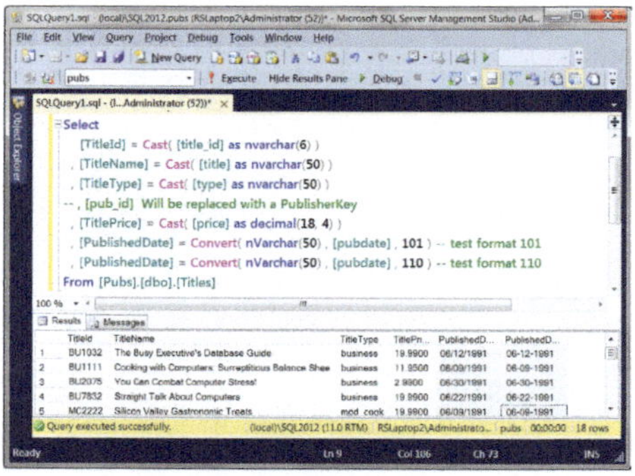

Figure 4.5. Schematic of adding more formatting options to the CONVERT() function.

Source: Natarjan, Creative Commons License.

The utilization of the CONVERT() functioning enables the transformation of DateTime types of data into a human-readable character format. However, given that the objective of the present ETL method does not involve the storage of dates in an alphabetical format, this functionality is deemed inconsequential. It is noteworthy that its existence is acknowledged, as it offers supplementary alternatives beyond the capabilities of the CAST() functions. It is important to note that the CAST() function is an established ANSI function, whereas the CONVERT() function is not. Consequently, if one requires to extract source data via tables within an Oracle or just MySQL database, utilizing CAST() remains the superior alternative (Astriani & Trisminingsih, 2016).

4.6.4. Searching for Substitute Key Values

As stated earlier, the inclusion of substitute keys within dimension tables is regarded as a recommended approach. The utilization of artificial columns facilitates the integration of data from disparate sources and enhances the efficacy of recording data modifications. It is imperative that the inter-table foreign key associations within our data storage facility are established on the substitute keys as opposed to the primary natural columns. These connections exhibit dissimilarities from the connections that have been established in the original database. Consequently, it is imperative to establish a method for retrieving the novel substitute key value by referencing the natural significant value within the primary tables (Bornea, Deligiannakis, Kotidis, & Vassalos, 2011). To facilitate a comprehensive comprehension of this matter, kindly refer to Figure 4.5.

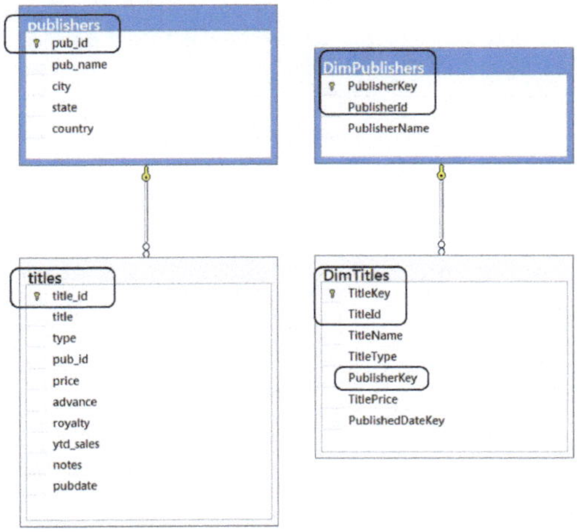

Figure 4.6. Image of tables with as well as without substitute keys are compared.

Source: Science Direct, Creative Commons License.

The figure illustrates that the initial connection among the titles as well as the creator's table within the OLTP file was established through the utilization of the pub_id column. The provision of links in the data storeroom now relies on the PublisherKey section, rendering the aforementioned statement invalid (Figure 4.6).

In order to populate the DimTitles table alongside accurate PublisherKey information, it is necessary to retrieve the new substitute key information of the publisher via referencing the initial relationship among publisher IDs which constituted the original typical keys. One feasible approach to achieve this objective is to generate a SQL join statement that retrieves data from the source as well as the destination tables based on their intrinsic key columns (Bala, Boussaid, & Alimazighi, 2017). The example presented in Listing 4–9 is illustrative.

Listing 4–9. the method of calculating a replacement key value through connecting the natural keys.

Select

[TitleId]=Cast([title_id] as nvarchar(4)), [TitleName]=Cast([title] as nvarchar(50))

, [TitleType]=Cast([type] as nvarchar(50))

, [PublisherKey]=[DWPubsSales].[dbo].[DimPublishers].[**PublisherKey**]

, [TitlePrice]=Cast([price] as decimal(18, 4))

, [PublishedDate]=[pubdate] From [**Pubs**].[dbo].[Titles]

Join [**DWPubsSales**].[dbo].[DimPublishers]

On [Pubs].[dbo].[Titles].[**pub_id**]=[DWPubsSales].[dbo].[DimPublishers].[**PublisherId**]

In order for the publisher's primary information to be accessible, it is necessary to populate the DimPublishers table beforehand. If a substitute key has not been generated, it will cease to exist (Pathania, Madan, Pandey, & Sharma, 2020). The following is a basic insert statement presented in Outlining 4–10 that achieves this objective.

Listing 4–10. The process of adding data to the DimPublishers table is referred to as inserting values

Insert Into [DWPubsSales].[dbo].[DimPublishers]

([PublisherId], [PublisherName])

Select

```
[PublisherId]=Cast([pub_id] as nchar(4))
[PublisherName]=Cast([pub_name] as nvarchar(50)) From [pubs].[dbo].[publishers]
```

4.6.5. Provide Conformity

Another frequently encountered ETL operation involves the task of standardizing data to enhance its readability and consistency. There exist two straightforward methodologies to achieve this objective through SQL code. The initial approach entails utilizing a SELECT-CASE statement, whereas the second approach involves the utilization of a lookup table (Azizi, Zamani, & Kolahdouz-Rahimi, 2017).

The creation of a SELECT-CASE statement involves the addition of the CASE part to any given statement. In Outlining 4–11, the values within the TitleType columns are extracted via the WHEN clause as well as subsequently matched against patterns such as "business," "mod_cook," "popular_comp," among others. This action is taken due to the inadequacy of the present values in terms of readability. For the sake of accuracy, it would have been appropriate to capitalize along with spell out the terms as "Modern Cooking" as well as "Popular Computing." Due to the lack of legibility in the abbreviations, it is necessary to implement a transformation process that adheres the data through a more comprehensible format (Ong et al., 2017). To accomplish this task, one must analyze a specific expression and subsequently contrast it with the desired pattern for matching. In the event that a specific pattern is matched by the expression stated within the WHEN clause, the outcome of the select command is converted into the outcome of the expression determined by the THEN clause (Yulianto, 2019). The process of achieving this form of transformation through the employment of the SQL SELECT-CASE assertion is demonstrated in Listing 4–11.

Listing 4–11. The process of aligning values can be achieved through the utilization of a SELECT-CASE construct.

```
Select
[TitleId]=Cast([title_id] as nvarchar(4)), [TitleName]=Cast([title] as nvarchar(50))
, [TitleType]=Case Cast([type] as nvarchar(50)) When 'business' Then 'Business'
When 'mod_cook' Then 'Modern Cooking'
When 'popular_comp' Then 'Popular Computing'
```

```
When 'psychology' Then 'Psychology'
When 'trad_cook' Then 'Traditional Cooking'
When 'UNDECIDED' Then 'Undecided'
End
, [PublisherKey]=[DWPubsSales].[dbo].[DimPublishers].[PublisherKey]
, [TitlePrice]=Cast([price] as decimal(18, 4))
, [PublishedDate]=[pubdate]
From [Pubs].[dbo].[Titles]
Join [DWPubsSales].[dbo].[DimPublishers]
On [Pubs].[dbo].[Titles].[pub_id]=[DWPubsSales].[dbo].[DimPublishers].[PublisherId]
```

Upon execution of the aforementioned code, the integers denoted as "mod_cook" undergo a conversion process resulting in the value "Modern Cooking," and so forth. The query SELECT is executed on each individual row within the table as well as is performed for each value contained within the column. Although efficacious, this approach has a drawback. In the event that one needs to apply identical transformations to multiple tables, it becomes necessary to replicate the SQL code for each respective table. Inconsistency in the capitalization of values used in different select statements can result in the creation of multiple renditions of identical data in your tables, for instance, "Modern Cooking" as well as "modern cooking" (Niinimäki & Niemi, 2009). The aforementioned title-type instances are improbable to be employed in any other part of the table, thus the likelihood of encountering an issue is minimal. Nevertheless, it is advisable to bear this in mind when adhering to naming conventions (da Silva, Times, & Kwakye, 2012). In the event that an individual intends to utilize identical transformation data for multiple tables, it may be advisable to consider employing the second alternative. The second alternative offers a mechanism for implementing these transformations through the generation of a lookup table, which is subsequently utilized to compare the entries of the initial table against those of the lookup table. Upon locating the match, retrieve the complied value via the searching table and substitute it for the initial value (Minton, 2004). One benefit of utilizing this alternative is its reusability, as it allows for multiple uses without the need for a repetitive definition of the roster of validated values. This approach is suitable in cases where a lookup table contains two-letter indicate abbreviations in certain cells and the complete name of the state in others (Thi & Nguyen, 2008). To obtain the full name of a state from its two-letter abbreviation within one or multiple tables, it is possible to utilize a lookup table as a means of achieving this objective. Utilizing a lookup table is a more suitable option in comparison to employing a Select-Case statement for data conformation, as indicated names are prone to manifest in numerous tables inside a specific data warehouse (Bansal, 2014). The creation of a query table and subsequent population with original as well as transformed values through the use of said table is demonstrated in Listing 4–12.

Listing 4–12. Aligning values through a reference table.

```
-- Create the lookup table
Create table [TitleTypeLookup] (
[TitleTypeKey] int Primary Key Identity
, [OriginalTitleType] nvarchar(50)
, [CleanTitleType] nvarchar(50)
)
-- Add the original and transformed data
Insert into [TitleTypeLookup]
([OriginalTitleType], [CleanTitleType])
Select
[OriginalTitleType]=[Type]
, [CleanTitleType]=Case Cast([type] as nvarchar(50))
When 'business' Then 'Business'
When 'mod_cook' Then 'Modern Cooking'
When 'popular_comp' Then 'Popular Computing'
When 'psychology' Then 'Psychology'
When 'trad_cook' Then 'Traditional Cooking'
When 'UNDECIDED' Then 'Undecided'
End
From [Pubs].[dbo].[Titles]
Group By [Type] -- get distinct values
-- Combine the data from the lookup table and the original table
Select
```

[TitleId]=Cast([title_id] as nvarchar(4)), [TitleName]=Cast([title] as nvarchar(50))

, [TitleType]=[CleanTitleType]

, [PublisherKey]=[DWPubsSales].[dbo].[DimPublishers].[PublisherKey]

, [TitlePrice]=Cast([price] as decimal(18, 4))

, [PublishedDate]=[pubdate]

From [Pubs].[dbo].[Titles]

Join [DWPubsSales].[dbo].[DimPublishers]

On [Pubs].[dbo].[Titles].[pub_id]=[DWPubsSales].[dbo].[DimPublishers].[PublisherId] **Join [DWPubsSales].[dbo].[TitleTypeLookup]**

On [Pubs].[dbo].[Titles].[type]=[DWPubsSales].[dbo].[TitleTypeLookup].[OriginalTitleType]

4.6.6. Generate Date Data

On certain occasions, the ETL procedure may not solely entail duplicating and altering pre-existing data, but rather, it may engender completely novel data. A common practice in data warehousing involves the utilization of a table that contains a sequential series of dates. This is typically achieved through the implementation of an INSERT assertion that is nested within a programmatic loop (Jörg & Deßloch, 2008).

Listing 4–14. Adding Entries to the DimDate Table.

-- Because the date table has no associated source table we can fill the data -- using a SQL script.

-- Create variables to hold the start and end date

DECLARE @StartDate datetime='01/01/1990' DECLARE @EndDate datetime='12/41/1995'

-- Use a while loop to add dates to the table

DECLARE @DateInProcess datetime SET @DateInProcess=@StartDate

WHILE @DateInProcess<= @EndDate

BEGIN -- Add a row into the date dimension table for this date **INSERT INTO DimDates** (

[Date]

, [DateName]

```
, [Month]
, [MonthName]
, [Quarter]
, [QuarterName]
, [Year]
, [YearName]
)
VALUES (
-- [Date]
@DateInProcess
-- [DateName]
, Convert(varchar(50), @DateInProcess, 110)+,' '
+ DateName(weekday, @DateInProcess)
-- [Month]
, Month(@DateInProcess)
-- [MonthName]
, Cast(Year(@DateInProcess) as varchar(4))+' – '
+ DateName(month, @DateInProcess) -- [Quarter]
, DateName(quarter, @DateInProcess)
-- [QuarterName]
, Cast(Year(@DateInProcess) as varchar(4))+' – '
+ 'Q'+DateName(quarter, @DateInProcess)
-- [Year]
, Year(@DateInProcess)
-- [YearName]
, Cast(Year(@DateInProcess) as Char(4))
)
-- Add a day and loop again
SET @DateInProcess=DateAdd(d, 1, @DateInProcess) END
-- Check the table SELECT Top 10 * FROM DimDates
```

It should be noted that SQL variables were utilized to denote the temporal scope of the dates, commencing on January 1st, 1990, and concluding on December 41st, 1995. It may appear peculiar, however, upon scrutinizing the dates provided by the Pubs database, it becomes evident that every single date falls within a specific range (Rodic & Baranovic, 2009) (Figure 4.7).

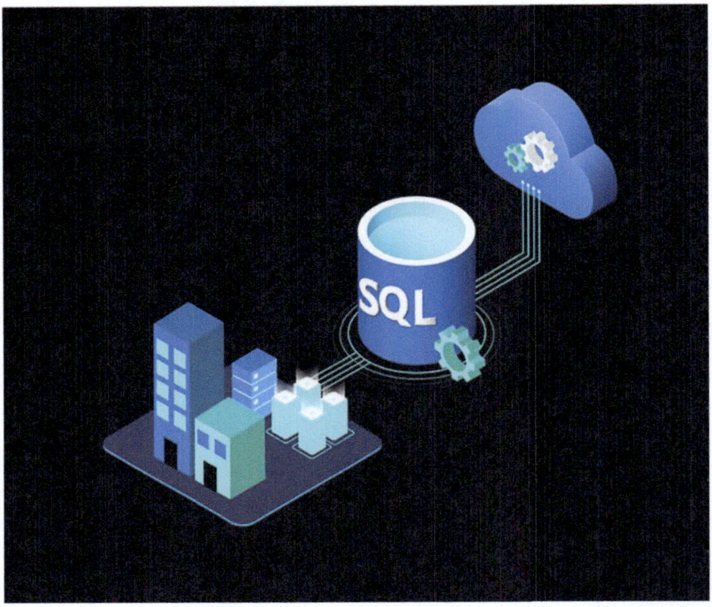

Figure 4.7. Illustration of importance of SQL.

Source: Microsoft, Creative Commons License.

4.6.7. Dealing with Nulls

The handling of null values within a database environment poses a unique challenge due to the existence of varying perspectives on the matter. Due to their inherent characteristics, null values offer several alternatives, and it is incumbent upon the user to determine the most effective strategy for a given scenario (Haase, Röseler, & Seidel, 2022). The concept of nulls is commonly associated with representing an unknown value. However, it is important to clarify the meaning of the term "unknown." Does this imply that the subject matter is inherently beyond comprehension and can never be acquired? Does this imply that the current status is uncertain, but there is a possibility of imminent availability? Does this imply that a certain value is not relevant or applicable in this specific instance? There are three possible scenarios that can be considered: firstly, the value might not be ascertainable; secondly, it may be absent; or thirdly, it might not be relevant.

Remember

When dealing with null values in the ETL process, consider using techniques like coalescing or conditional logic to handle nulls appropriately and maintain data .integrity

Due to the inherent confusion of the expression "null," it is imperative to determine the most reliable interpretation to avoid protracted debates regarding the veracity of one's reports. The following discourse aims to analyze each available alternative in order to facilitate the decision-making process (El Akkaoui & Zimányi, 2009).

4.6.8. Nulls in a Fact Table

The presence of null values within the fact table may pertain to either the dimension-related keys or just the values that were measured. It is advisable to retain the null value if it represents a quantifiable parameter, as it may signify an absence of information or an undeveloped state. The reason for this is that the majority of database products possess the capability to effectively consolidate null values into overall and partial sums. The substitution of values, including zero, for null can result in erroneous calculations (de S. Ribeiro, Goldschmidt, & Cavalcanti, 2011).

To gain comprehension of this matter, please refer to Listing 4–14. The instance demonstrates the disparity that arises when employing the cumulative function AVERAGE().

Listing 4–14. The present piece of writing examines the manner in which null values operate in conjunction with collected functions.

```
-- Using a null you get the correct answer of 4
Select Avg([Amt]) From (
Select [Amt] = 2 Union
Select [Amt] = 4
Union
Select [Amt] = null
) as aDemoTableWithNull
-- Using a zero you get the incorrect answer of 2
```

```
Select Avg([Amt]) From (
Select [Amt] = 2 Union
Select [Amt] = 4
Union
Select [Amt] = 0
) as aDemoTableWithZero
```

When encountering a null value within the fact table which pertains to a dimensional key, it is recommended to generate a synthetic key value that can provide additional context to the null. It should be noted that multiple artificial keys may be required to represent each instance of null values. It is common for programmers to seek to cluster null data and attribute implicit connotations to such null values. Whilst this methodology proves to be advantageous in ascertaining the count of unshipped orders, it is not without its drawbacks (Andersen, Thomsen, & Torp, 2018). What are the consequences of an incorrect implicit significance?" If a package has been dispatched without the corresponding shipping date being documented, there exists the possibility of the package being dispatched a second time. In order to address these concerns, it is recommended to offer methods for substituting null values with a collection of descriptive dimensional characteristics. Subsequently, a SELECT-CASE searching transformation can be employed, akin to the one expounded upon within the "Provide Conformity" segment of this part, to substitute the null values with the aforementioned evocative dimensional values (Bindal & Khurana, 2015).

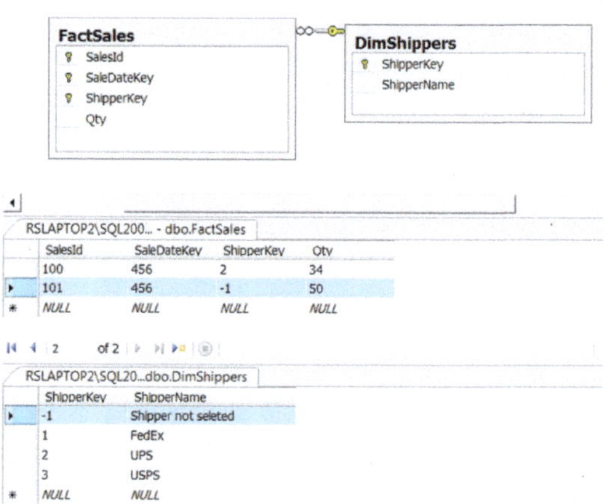

Figure 4.8. The utilization of a substitute key in the presence of null values.

Source: Science direct, Creative Commons License.

The following is an illustration of a basic null lookup: in the event that a transaction record within the fact table lacks a corresponding ShipperId due to the absence of a selection, a ShipperId using a name column value of "Shipper not selected" can be utilized, as depicted in Figure 4.8.

4.6.9. Nulls in a Dimension Table

It is not recommended to retain null values within dimensional tables. Alternatively, they may be omitted or modified. One straightforward approach to managing null values is to prohibit their inclusion within the data warehouses. The concept entails that despite the null values being potentially reportable within the OLTP surroundings, their lack of specificity renders them unsuitable for utilization within the data warehouses. In instances when it is deemed appropriate to prohibit nulls, the matter can be readily addressed, as exemplified in Outlining 4–15.

Listing 4–15. Excluding Nulls with the Where Clause

Select

[TitleId]=Cast([title_id] as nvarchar(4))…, [TitleName]=Cast([title] as nvarchar(50))

, [TitleType]=Cast([type] as nvarchar(50))

, [PublisherKey]=[DWPubsSales].[dbo].[DimPublishers].[PublisherKey]

, [TitlePrice]=Cast([price] as decimal(18, 4))

, [PublishedDate]=[pubdate]

From [Pubs].[dbo].[Titles]

Join [DWPubsSales].[dbo].[DimPublishers]

On [Pubs].[dbo].[Titles].[pub_id]=[DWPubsSales].[dbo].[DimPublishers].[PublisherId]

Where [Pubs].[dbo].[Titles].[Title_Id] **Is Not Null And** [Pubs].[dbo].[Titles].[Title] **Is Not Null And** [Pubs].[dbo].[Titles].[Type] **Is Not Null**

And [Pubs].[dbo].[Titles].[Price] **Is Not Null**

And [Pubs].[dbo].[Titles].[PubDate] **Is Not Null**

While this methodology may be effective for certain tables, it possesses the capacity to omit a substantial amount of pertinent data required for a comprehensive BI resolution. Whenever feasible, it is preferable to convert null values through descriptive values for enhanced precision (Roblero, Jácome, & Argotte, 2012). An approach to achieve this objective is by utilizing the ISNULL() functioning. The ISNULL() operation, as demonstrated in Listing 4–14, facilitates the conversion of null values to the term "Unknown." The term "Unknown" carries a higher degree of descriptiveness compared to the utilization of "null," albeit the difference may be considered marginal.

Listing 4–14. Converting Null Values with the ISNULL() Function

Select

[TitleId]=Cast(**isNull([title_id], -1)** as nvarchar(4))

, [TitleName]=Cast(**isNull([title], 'Unknown')** as nvarchar(50))

, [TitleType]=Cast(**isNull([type], 'Unknown')** as nvarchar(50))

, [PublisherKey]=[DWPubsSales].[dbo].[DimPublishers].[PublisherKey]

, [TitlePrice]=Cast(**isNull([price], -1)** as decimal(18, 4))

, [PublishedDate]=**isNull([pubdate], '01/01/1900')**

From [Pubs].[dbo].[Titles]

Join [DWPubsSales].[dbo].[DimPublishers]

On [Pubs].[dbo].[Titles].[pub_id]=[DWPubsSales].[dbo].[DimPublishers].[PublisherId]

The insertion of the string value "Unknown" into columns containing non-character data types, including integers as well as datetime, isn't permissible due to the presence of null values. Thus, the ISNULL() function may be employed to transform the date of publication to a format such as 01/01/1900. The calendar year 1900 serves as an initial reference for numerous date data types along with is improbable to signify an actual occurrence, particularly within the confines of this specific database. This implies that the absence of a specific publication date may be inferred using this criterion without any bearing on its actual release date. It is important to note that utilizing this date would not be suitable if the dataset incorporates this specific date as a valid value (Andrews, 2015).

4.6.10. A Null Lookup Table

While several conventional methods for handling null values have become common in data stores, an alternative approach involves utilizing reference search values from the option of a dedicated search table or pre-existing dimensional tables. Search tables, also referred to by the term domain tables, are supplementary tables that aren't integrated into the dimension model

(Tziovara, Vassiliadis, & Simitsis, 2007). Rather, they contain descriptive values that may be utilized as references within the dimensional tables.

In order for the search table to operate effectively, it is imperative to ascertain the significance of every null value present in the original data and subsequently convert said null values into the search ID that can be utilized for cross-referencing with the query table (Sharma & Attar, 2016).

The challenge associated with utilizing a dedicated searching table is that every dimension containing null values results in the formation of a snowflake schema. While not necessarily presenting a significant issue, this contravenes the principle of simplicity. Frequently, it is possible to integrate a searchable database into the dimension tables (Hänel & Felden, 2017). In instances where a date column within a fact or dimensions table is null, it is possible to utilize a dimension table, including the DimDates table, just like a lookup table. This is contingent upon the presence of rows within the table that include relevant lookup data (Desarkar, 2004). The aforementioned lookup rows comprise descriptive values which signify null interpretations. The DimDates table can be seen to have two lookup rows added in Listing 4–17.

Listing 4–17. The procedure of integrating additional reference values with the DimDates dimension table.

```
Set Identity_Insert [DimDates] On
INSERT INTO [DWPubsSales].[dbo].[DimDates] ([DateKey] -- This is
normally added automatically
, [Date]
, [DateName]
, [Month]
, [MonthName]
, [Quarter]
, [QuarterName]
, [Year]
, [YearName]
)
VALUES
(-1 -- This will be the Primary key for the first lookup value
, '01/01/1900'
, 'Unknown Day'
, -1, 'Unknown Month'
, -1
```

, 'Unknown Quarter'
, -1
, 'Unknown Year'
)
, -- add a second row
(-2 -- This will be the Primary key for the second lookup value
, '02/01/1900'
, 'Corrupt Day'
, -2
, 'Corrupt Month'
, -2
, 'Corrupt Quarter'
, -2
, 'Corrupt Year'
)

Set Identity_Insert [DimDates] Off

The DimDates table was updated with rows for dates in Outlining 4–14, however, null search values were not incorporated. The IDENTITY choice, by default, restricts the manual insertion of values onto the column. In the event that one desires to manually input a value, it is possible to do so by enabling SQL's IDENTITY_INSERT option through the utilization of the SET IDENTITY_INSERT <table name> ON SQL command (Qu & Deßloch, 2017). Upon completion of the aforementioned task, supplementary lookup values may be incorporated into the dimensions table.

After incorporating the additional lookup rows, they can be utilized in the remaining tables. The code presented in Listing 4–18 is capable of achieving this task. The implementation employs an outer join operation to incorporate all values present in the DimTitles, regardless of cases where no corresponding date is detected within the DimDates table. In instances where a corresponding date is not present on a given row, the outer link operation mandates the inclusion of a null value within the output (Awiti & Zimányi, 2019). Hence, the ISNULL() function may be employed to transform the recently generated null parameter to -1, which is subsequently correlated with the Undetermined date within the DimDates table.

Listing 4–18. The process of cross-referencing with the dimensional keys of the DimDates table.

Select
[TitleId]=Cast(isNull([title_id], -1) as nvarchar(4))

```
, [TitleName]=Cast(isNull([title], 'Unknown') as nvarchar(50))
, [TitleType]=Cast(isNull([type], 'Unknown') as nvarchar(50))
, [PublisherKey]=[DWPubsSales].[dbo].[DimPublishers].[PublisherKey]
, [TitlePrice]=Cast(isNull([price], -1) as decimal(18, 4))
, [PublishedDateKey]=isNull([DWPubsSales].[dbo].[DimDates].
[DateKey], -1)
From [Pubs].[dbo].[Titles]
Join [DWPubsSales].[dbo].[DimPublishers]
On [Pubs].[dbo].[Titles].[pub_id]=[DWPubsSales].[dbo].[DimPublishers].
[PublisherId]
Left Join [DWPubsSales].[dbo].[DimDates] -- The "Left" keeps dates not
found in DimDates
On [Pubs].[dbo].[Titles].[pubdate]=[DWPubsSales].[dbo].[DimDates].
[Date]
```

4.7. THE SQL QUERY DESIGNER

As evident, the SQL code has the potential to rapidly become intricate. One can facilitate the process of coding by employing the SQL Query Designers that is included in either SQL Management SSIS. Both of these instruments are almost indistinguishable and generate SQL code through a graphical user interface (GUI) (Kießling & Köstler, 2002).

The Query Designers integrated into SQL Management Studio facilitate the creation of multiple statements within a singular SQL script file, while the Query Designer incorporated into the SSIS framework for development lacks this capability. Hence, it is probable that the Query Designers of Management Studio would be deemed as the more potent alternative (Prior & Lister, 2004).

To initiate the Query Designer of SQL Server Management Studio, one must launch a fresh query window as well as concurrently tap the Ctrl+Shift+Q keys. The designer can be accessed via the queries menu located at the top level of Management Studios or via selecting the Designing a Query within Editor choice from the contextual menu that appears upon right-clicking the empty portion of a query window. It is recommended that the final option be utilized. Typically, the initial step involves appending a remark within the query display to specify the intended objective (Kearns, Shead, & Fekete, 1997). Subsequently, a novel line is generated beneath the commentary, as well as a right-click upon the vacant area is executed to gain entry to the contextual menu (refer to Figure 4.8).

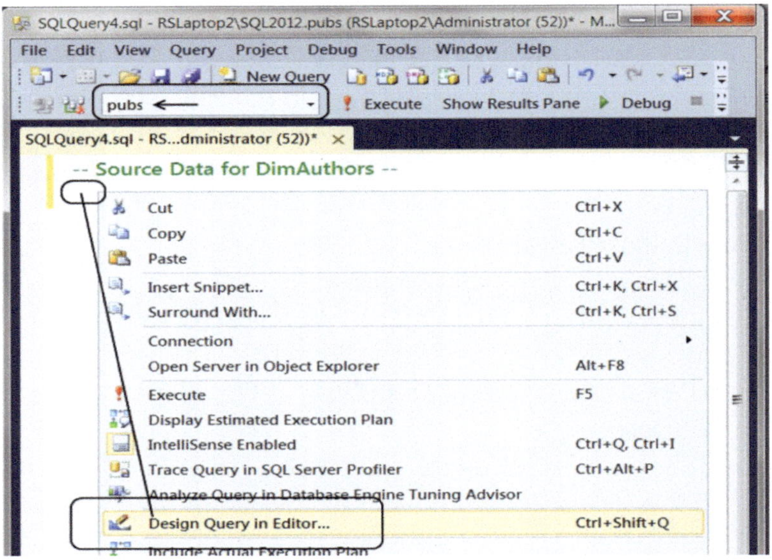

Figure 4.9. Image of initiating the query editor.

Source: Science direct, Creative Commons License.

Upon launching the Query Designer, the user will be presented with the Add Table pop-up window as well as an extensive directory of database tables. Upon choosing one or multiple tables within this interface, the names of said tables will be incorporated into the FROM section within the SQL SELECT query that is presently being generated (Date, 1989). Upon selecting the Authors table as well as subsequently clicking the Add button, the creator's table is going to be appended to the query window, as depicted in Figure 4.9. To incorporate multiple tables into a query, users can continue holding down the Ctrl key, select the desired tables, and subsequently click on the Add button. This functionality enables the creation of intricate queries like those involving SQL joins. Upon selecting all the desired tables, the user should proceed by clicking the Close button so as to terminate the Add Tables pop-up window (Rankins, Jensen, & Bertucci, 2003).

Upon closure within the Add Tables pop-up window, the tables will be displayed at the uppermost section of the Query Designers window, while the SQL code utilized to query each table will be situated at the lowermost section.

It is necessary to extract data that comes from the initial table of the original writer. Consequently, a SQL SELECT clause is composed in order to retrieve the aforementioned data. By selecting the checkboxes corresponding to the desired columns, the Query Designers will incorporate them into the SELECT statement being constructed. There exist several discrepancies between the Authors as well as DimAuthors tables. Consequently, it is necessary to implement several modifications to the fundamental query. The operations encompassed in this context comprise the amalgamation of data from various columns into a singular column of facts, the alteration of the names of the extant columns, as well as the transformation of the data depending on each column (Albrecht & Naumann, 2008).

Frequent data manipulations include merging 2 columns to generate a unified name and altering the data kind of the original data. The objective of the present example is to merge the initial and final names of a writer into a solitary column. It is generally accepted to maintain separate names within an OLTP database. However, it may not be beneficial in the context of our sales data storage, unless there is a possibility of generating reports that require aggregation of measures based on an individual's first or final name. It is assumed that a sales report supplying details about the number of sales according to the author's maiden name is unnecessary within the DWPubsSales data center. Thus, the amalgamation of an author's initial and surname is proposed as a means to streamline the reporting procedure. To amalgamate this information, a straightforward concatenation can be incorporated into the query (Skoutas & Simitsis, 2006).

The act of renaming columns constitutes a basic transformation. The process of achieving this conversion involves the inclusion of the alias into the Alias column located in the Query designers.

As demonstrated earlier, the modification of data types may be achieved through the utilization of a combination of the CAST() as well as CONVERT() functioning. These functions can be directly incorporated through the query display to generate an alternative SQL expression. Regrettably, the Query Designers tool isn't meant for this purpose and may misinterpret the user's intentions despite the utilization of valid SQL syntax. Upon navigating a way through the Column cell during which the expression has been inputted, a message of error is going to be displayed indicating that the transformation may be deemed superfluous. It is possible to disregard the warning message provided that the syntax used in SQL is deemed accurate and suitable for the intended task. However, it is necessary to modify the code despite the need to circumvent the notification of error (Bao, Song, Leng, Wang, & Yu, 2007).

The code that has been formatted will resemble the code presented in Listing 4–19. It should be noted that an additional N has been incorporated into the process of data conversion. The symbol 'N' denotes that the interval among the two apostrophes must be interpreted as Unicode. From a technical standpoint, there are no inherent issues with this approach. However, it could potentially cause confusion for individuals who are tasked with developing the corresponding code. The utilization of the N symbol has been extensively documented and may be readily researched to ascertain its significance. However, it is frequently misconstrued by novice SQL programmers (Atigui, Ravat, Teste, & Zurfluh, 2012).

Listing 4–19. A code snippet that can be modified by the Query Designers without explicit authorization.

SELECT au_id AS AuthorId, **CAST((au_fname+N' '+au_lname)** AS nVarchar(100)) AS AuthorName, state AS AuthorState

FROM authors

An additional concern pertaining to the Query Designers is its code formatting methodology. As stated in the introductory section of this section, maintaining uniform formatting in a person's code is indicative of a professional approach. When attempting to format the source code within the Query Designers window, the system will disregard the user's input and instead modify the code according to what it deems to be optimal. The Query Designers are believed to possess a sense of self-importance. It is possible to perform code as well as conversion code maintenance subsequent to the closure of the Query Designers. However, failure to utilize this functionality may result in a vexing experience (Stumptner, Freudenthaler, & Krenn, 2012). The reformatting of the code generated through the Query Designer. Furthermore, data conversion has been incorporated into the state column subsequent to the closure of the Query designers, with the intention of circumventing the recurrence of the error.

4.8. UPDATING BI DOCUMENTATION

The significance of efficient documentation cannot be overstated in the context of BI initiatives, especially in relation to SQL-based ETL (Extract, Transformation, Load) processing. Documentation functions as a point of reference for various stakeholders, including developers, analysts, as well as other relevant parties, with the aim of guaranteeing the BI system's coherence, comprehensibility, and sustainability (Ferrari & Russo, 2016).

It is imperative to maintain up-to-date documentation in order to reflect any modifications made within the ETL process as it progresses. The present chapter centers on optimal methodologies for revising BI documentation, alongside a specific emphasis on ETL procedures utilizing SQL.

1. Documentation regarding ETL adjustments is crucial to maintain alignment with the present version of the system. This involves updating the documentation whenever modifications occur to the ETL method, such as the addition of fresh data sources, adjustments to data alterations, or shifts to load processes. The following are a set of instructions to adhere to:
   a. Identification of Changes: Conduct an analysis of the alterations carried out on the ETL procedure as well as ascertain the particular domains that necessitate a revision in the records.
   b. In the event that additional data sources have been included, it is imperative to furnish comprehensive information regarding their nomenclature, descriptions, geographical locations, and pertinent connection details (Ghilic-Micu, Mircea, & Stoica, 2010).
   c. In this section, the data transformations will be explicated. The modifications applied to the data modifications will be recorded, encompassing the reasoning, equations, or SQL queries employed for the alteration of the data. Revise the records to incorporate the revised regulations for transformation.
   d. To maintain accuracy and consistency, it is imperative to update the import procedures records in accordance with any modifications made to the methods. This will ensure that any updated actions involved in transferring the data through the system being targeted are accurately reflected in the records (Negash, 2004).
   e. In light of the modifications made to the ETL process, it is necessary to specify any additional dependencies which have arisen. It is imperative to establish an in-depth comprehension of the entire process by documenting the interrelationships between various data sources, modifications, as well as load procedures (Mundy & Thornthwaite, 2011).
2. The practice of maintaining control over revisions as well as versioning is crucial in documentation management as it facilitates change tracking as well as enables retrieval of previous versions when necessary. It is advisable to contemplate the adoption of a modification control system, like Git, as a means of overseeing

your documentation. This feature enables the user to monitor the chronology of modifications, examine discrepancies among iterations, and restore previous versions in case of need (Muller & Hart, 2016).

3. The process of updating BI documentation encompasses not only internal modifications but also necessitates effective communication of said changes to pertinent stakeholders. This practice aids in achieving a shared understanding among all parties involved and comprehending the consequences of any alterations made. Please take into account the following:
   a. Disseminate Notification: Disseminate notification to all teammates who participate in the method of ETL regarding the updates made to the documentation. Effective communication may be facilitated through various means such as team meetings, electronic mails, or designated communication channels.
   b. Conducting training sessions is recommended in the event of substantial modifications to the ETL process. This will aid in acquainting everyone involved with the revised documentation. This measure guarantees a comprehensive comprehension of the novel protocols and enables their efficient adherence by all individuals involved (Goul, Marjanovic, Baxley, & Vizecky, 2012).
   c. In the event that the BI system has other individuals who depend on it, it is imperative to guarantee that any modifications made to the ETL method are properly documented in the client's documentation. This entails the revision of user manuals, frequently asked questions (FAQs), and other pertinent resources to ensure that users are apprised of the modifications (El-Gayar & Timsina, 2014).

4. The process of updating BI documentation is a continuous task that involves maintenance and regular review. It is imperative to allot sufficient time for periodic evaluations and upkeep in order to guarantee the precision and currency of the documentation. The following practices are worth considering:

a. Allocate ownership: Nominate an accountable individual or group to uphold the documentation. The implementation of accountability measures aids in maintaining the accuracy and currency of documentation (Luhn, 1958).

b. Establishing periodic reviews is recommended to detect any deficiencies or obsolete information within the documentation. The periodicity of performing this task may be determined on a quarterly or semi-annual basis, or alternatively, may be contingent upon the regularity of alterations to the ETL method.

c. Cultivate feedback by prompting team members to provide their input on the documentation. It is possible that they possess useful knowledge or suggestions that could enhance the clarity, completeness, as well as organization of the subject matter.

It is imperative to update the documentation of your BI to guarantee that it precisely represents the present condition of your ETL processing with Structured Query Language (SQL). By adhering to the recommended guidelines presented in this chapter, one can ensure the maintenance of thorough and current documentation that serves the interests of all parties associated with the BI project. To attain enduring success in one's BI endeavors, it is imperative to document alterations, execute revision control, proficiently communicate updates, as well as consistently review and uphold one's documentation (Sherman, 2014).

Activity 1.1

As a data engineer, you have been assigned to develop an ETL process for a retail company's sales data. Outline the steps you would take to perform the ETL programming, including the choice between full loading and loading in stages. Discuss the importance of data reduction techniques, column aliases, and handling null values in the transformation algorithm. Additionally, explain the role of the .SQL Query Designer in designing and executing efficient ETL queries

4.9. SUMMARY

Chapter four provides an in-depth understanding of the ETL process with a focus on using SQL for data integration and transformation. It begins with an introduction to the ETL programming and explores the considerations between full loading and loading in stages. The chapter delves into the crucial aspect of transforming logic identification, emphasizing the development of a transformation algorithm. It covers various transformation techniques, such as data reduction, column aliases for better readability, data types conversion, searching for substitute key values, providing conformity, generating date data, and dealing with null values in both fact and dimension tables. The chapter also introduces the SQL Query Designer, which plays a significant role in designing and executing efficient ETL queries. It emphasizes the importance of updating BI documentation to reflect the changes made during the ETL process, ensuring comprehensive and up-to-date documentation. Overall, this chapter provides valuable insights into the ETL process using SQL, highlighting the importance of data integration and transformation in modern data management. It equips readers with the knowledge and techniques needed to perform efficient ETL programming, make informed choices in the ETL process, and handle various data scenarios effectively.

Review Questions

1. Explain the purpose and significance of the ETL process in data management.
2. Compare and contrast full loading and loading in stages approaches in the ETL process.
3. What is data reduction, and how does it contribute to storage optimization?
4. Describe the use of column aliases in SQL transformations and its benefits.
5. Discuss the importance of data types conversion in the ETL process.
6. How can you search for substitute key values during the transformation phase?
7. Explain the concept of providing conformity in data transformation.
8. Discuss the generation of date data in the ETL process and its significance.
9. How can null values be handled effectively in the ETL process?
10. Explain the implications of nulls in fact tables and dimension tables in a data warehouse environment.

REFERENCES

1. Albrecht, A., & Naumann, F., (2008). Managing ETL processes. *NTII, 8*, 12–15.
2. Andersen, O., Thomsen, C., & Torp, K., (2018). *SimpleETL: ETL Processing by Simple Specifications.* Paper presented at the proceedings of the 20th international workshop on design, optimization, languages, and analytical processing of big data co-located with 10th EDBT/ICDT Joint Conference.
3. Andrews, M., (2015). NoETL: ETL code generation for a dimensional-data warehouse. *VMWare Technical Journal, 4*(1), 41–45.
4. Astriani, W., & Trisminingsih, R., (2016). Extraction, transformation, and loading (ETL) module for hotspot spatial data warehouse using GeoKettle. *Procedia Environmental Sciences, 33*, 626–634.
5. Atigui, F., Ravat, F., Teste, O., & Zurfluh, G., (2012). *Using OCL for Automatically Producing Multidimensional Models and ETL Processes.* Paper presented at the Data Warehousing and Knowledge Discovery: 14th International Conference, DaWaK 2012, Vienna, Austria, September 3–6, 2012. Proceedings 14.
6. Awiti, J., & Zimányi, E., (2019). *An XML Interchange Format for ETL Models.* Paper presented at the new trends in databases and information systeMS: ADBIS 2019 Short Papers, Workshops BBIGAP, QAUCA, SemBDM, SIMPDA, M2P, MADEISD, and Doctoral Consortium, Bled, Slovenia, September 8–11, 2019, Proceedings 23.
7. Awiti, J., Vaisman, A. A., & Zimányi, E., (2020). Design and implementation of ETL processes using BPMN and relational algebra. *Data & Knowledge Engineering, 129*, 101837.
8. Azizi, B., Zamani, B., & Kolahdouz-Rahimi, S., (2017). *Contract Verification of ETL Transformations.* Paper presented at the 2017 7th International Conference on Computer and Knowledge Engineering (ICCKE).
9. Bala, M., Boussaid, O., & Alimazighi, Z., (2014). *P-ETL: Parallel-ETL Based on the MapReduce Paradigm.* Paper presented at the 2014 IEEE/ACS 11th International Conference on Computer Systems and Applications (AICCSA).
10. Bala, M., Boussaid, O., & Alimazighi, Z., (2017). A fine-grained distribution approach for ETL processes in big data environments. *Data & Knowledge Engineering, 111*, 114–136.

11. Bansal, S. K., (2014). *Towards a Semantic Extract-Transform-Load (ETL) Framework for Big Data Integration.* Paper presented at the 2014 IEEE International Congress on Big Data.
12. Bao, Y., Song, J., Leng, F., Wang, D., & Yu, G., (2007). Study and implementation of a new SQL-based ETL approach. *Wuhan University Journal of Natural Sciences, 12*(5), 804–808.
13. Bindal, P., & Khurana, P., (2015). Etl life cycle. *International Journal of Computer Science and Information Technologies.* Descargado de http://bit. ly/2UXaXX6.
14. Bornea, M. A., Deligiannakis, A., Kotidis, Y., & Vassalos, V., (2011). *Semi-streamed Index Join for Near-Real Time Execution of ETL Transformations.* Paper presented at the 2011 IEEE 27th International Conference on Data Engineering.
15. Bouman, R., Casters, M., & Van, D. J., (2010). *Pentaho Kettle Solutions: Building Open Source ETL Solutions with Pentaho Data Integration.* John Wiley & Sons.
16. Da Silva, M. S., Times, V. C., & Kwakye, M. M., (2012). A framework for ETL systems development. *Journal of Information and Data Management, 3*(3), 300.
17. Date, C. J., (1989). *A Guide to the SQL Standard.* Addison-Wesley Longman Publishing Co., Inc.
18. De Ribeiro, L. S., Goldschmidt, R. R., & Cavalcanti, M. C., (2011). *Complementing Data in the ETL Process.* Paper presented at the Data Warehousing and Knowledge Discovery: 13th International Conference, DaWaK 2011, Toulouse, France, August 29-September 2, 2011. Proceedings 13.
19. Desarkar, M. S., (2004). *Data Profiling for ETL Processes.* Indian Institute of Technology, Kanpur, India.
20. Diouf, P. S., Boly, A., & Ndiaye, S., (2017). *Performance of the ETL Processes in Terms of Volume and Velocity in the Cloud: State of the Art.* Paper presented at the 2017 4th IEEE International Conference on Engineering Technologies and Applied Sciences (ICETAS).
21. El Akkaoui, Z., & Zimányi, E., (2009). *Defining ETL Worfklows Using BPMN and BPEL.* Paper presented at the Proceedings of the ACM twelfth international workshop on Data warehousing and OLAP.
22. El-Gayar, O., & Timsina, P., (2014). *Opportunities for Business Intelligence and Big Data Analytics in Evidence Based Medicine.*

Paper presented at the 2014 47th Hawaii International Conference on System Sciences.

23. El-Sappagh, S. H. A., Hendawi, A. M. A., & El Bastawissy, A. H., (2011). A proposed model for data warehouse ETL processes. *Journal of King Saud University-Computer and Information Sciences, 23*(2), 91–104.
24. Ferrari, A., & Russo, M., (2016). *Introducing Microsoft Power BI.* Microsoft Press.
25. Galici, R., Ordile, L., Marchesi, M., Pinna, A., & Tonelli, R., (2020). Applying the ETL process to blockchain data. prospect and findings. *Information, 11*(4), 204.
26. Ghilic-Micu, B., Mircea, M., & Stoica, M., (2010). The audit of business intelligence solutions. *Informatica Economica, 14*(1), 66.
27. Goul, M., Marjanovic, O., Baxley, S., & Vizecky, K., (2012). *Managing the Enterprise Business Intelligence App Store: Sentiment Analysis Supported Requirements Engineering.* Paper presented at the 2012 45th Hawaii international conference on system sciences.
28. Gour, V., Sarangdevot, S., Tanwar, G. S., & Sharma, A., (2010). Improve performance of extract, transform and load (ETL) in data warehouse. *International Journal on Computer Science and Engineering, 2*(3), 786–789.
29. Haase, C., Röseler, T., & Seidel, M., (2022). *METL–a Modern ETL Pipeline with a Dynamic Mapping Matrix.* arXiv preprint arXiv:2203.10289.
30. Hänel, T., & Felden, C., (2017). Design and evaluation of an analytical framework to analyze and control production processes. *Procedia CIRP, 62*, 141–146.
31. Jain, N., Liao, G., & Willke, T. L., (2013). *GraphBuilder: Scalable Graph ETL Framework.* Paper presented at the First international workshop on graph data management experiences and systems.
32. Jörg, T., & Deßloch, S., (2008). *Towards Generating ETL Processes for Incremental Loading.* Paper presented at the Proceedings of the 2008 international symposium on database engineering & applications.
33. Kakish, K., & Kraft, T. A., (2012). *ETL Evolution for Real-Time Data Warehousing.* Paper presented at the Proceedings of the Conference on Information Systems Applied Research ISSN.

34. Karakasidis, A., Vassiliadis, P., & Pitoura, E., (2005). *ETL Queues for Active Data Warehousing.* Paper presented at the Proceedings of the 2nd international workshop on Information quality in information systems.
35. Kearns, R., Shead, S., & Fekete, A., (1997). *A Teaching System for SQL.* Paper presented at the Proceedings of the 2nd Australasian conference on Computer science education.
36. Kießling, W., & Köstler, G., (2002). *Preference SQL—Design, Implementation, Experiences.* Paper presented at the VLDB'02: Proceedings of the 28th International Conference on Very Large Databases.
37. Luhn, H. P., (1958). A business intelligence system. *IBM Journal of Research and Development, 2*(4), 314–319.
38. Minton, P., (2004). The conformed ETL architecture; pounding square pegs into round holes (without cutting corners). *Information Management, 14*(9), 8.
39. Muller, L., & Hart, M., (2016). *Updating Business Intelligence and Analytics Maturity Models for New Developments.* Paper presented at the Decision Support Systems VI-Addressing Sustainability and Societal Challenges: 2nd International Conference, ICDSST 2016, Plymouth, UK, May 23–25, 2016, Proceedings 2.
40. Mundy, J., & Thornthwaite, W., (2011). *The Microsoft Data Warehouse Toolkit: With SQL Server 2008 R2 and the Microsoft Business Intelligence Toolset*: John Wiley & Sons.
41. Mutanu, L., & Khan, Z., (2021). Enhancing the QoS of a data warehouse through an improved ETL approach. *Journal of Language, Technology & Entrepreneurship in Africa, 12*(2), 89–102.
42. Natarajan, S., (2011). *Enhancing Web ETL Tool.* California State University, Sacramento.
43. Negash, S., (2004). Business intelligence. *Communications of the Association for Information Systems, 13*(1), 15.
44. Niinimäki, M., & Niemi, T., (2009). An ETL process for OLAP using RDF/OWL ontologies. *Journal on Data Semantics XIII*, 97–119.
45. Ong, T. C., Kahn, M. G., Kwan, B. M., Yamashita, T., Brandt, E., Hosokawa, P., & Schilling, L. M., (2017). Dynamic-ETL: A hybrid approach for health data extraction, transformation, and loading. *BMC Medical Informatics and Decision Making, 17*(1), 1–12.

46. Pathania, A., Madan, J., Pandey, R., & Sharma, R., (2020). Effect of structural and temperature variations on perovskite/Mg 2 Si based monolithic tandem solar cell structure. *Applied Physics A, 126*, 1–12.

47. Perron, M., Castro, F. R., DeWitt, D., & Madden, S., (2020). *Starling: A Scalable Query Engine on Cloud Functions.* Paper presented at the Proceedings of the 2020 ACM SIGMOD International Conference on Management of Data.

48. Prasath, N., & Sreemathy, J., (2021). *A New Approach for Cloud Data Migration Technique Using Talend ETL Tool.* Paper presented at the 2021 7th International Conference on Advanced Computing and Communication Systems (ICACCS).

49. Prior, J. C., & Lister, R., (2004). The backwash effect on SQL skills grading. *ACM SIGCSE Bulletin, 36*(3), 32–36.

50. Qu, W., & Deßloch, S., (2017). Incremental ETL pipeline scheduling for near real-time data warehouses. *Datenbanksysteme* für Business, *Technologie und Web (BTW 2017).*

51. Rankins, R., Jensen, P., & Bertucci, P., (2003). *Microsoft SQL Server 2000 Unleashed.* Sams Publishing.

52. Roblero, J., Jácome, N., & Argotte, L., (2012). *Important Considerations for an ETL Process in CFE's Health and Security Area.* Paper presented at the Proceedings of the World Congress on Engineering and Computer Science.

53. Rodic, J., & Baranovic, M., (2009). *Generating Data Quality Rules and Integration Into ETL Process.* Paper presented at the Proceedings of the ACM twelfth international workshop on Data warehousing and OLAP.

54. Rodzi, N. A. H. M., Othman, M. S., & Yusuf, L. M., (2015). *Significance of Data Integration and ETL in Business Intelligence Framework for Higher Education.* Paper presented at the 2015 International Conference on Science in Information Technology (ICSITech).

55. Sabtu, A., Mohd, A. N. F., Amir S. N. N., Adli, I. S., Mohd, Y. O., Sarkan, H., & Chuprat, S., (2017). The challenges of extract, transform and load (ETL) for data integration in near Realtime environment. *Journal of Theoretical & Applied Information Technology, 95*(22).

56. Seenivasan, D., (2023). Exploring popular ETL testing techniques. *International Journal of Computer Trends and Technology, 71*(2), 32–39.

57. Sharma, K., & Attar, V., (2016). *Generalized Big Data Test Framework for ETL Migration.* Paper presented at the 2016 International Conference on Computing, Analytics and Security Trends (CAST).
58. Sherman, R., (2014). *Business Intelligence Guidebook: From Data Integration to Analytics.* Newnes.
59. Simitsis, A., (2004). *Modeling and Optimization of Extraction-Transformation-Loading (ETL) Processes in Data Warehouse Environments.* National Technical University of Athens: PhD Thesis, Athens, Greece.
60. Skoutas, D., & Simitsis, A., (2006). *Designing ETL Processes Using Semantic Web Technologies.* Paper presented at the Proceedings of the 9th ACM international workshop on Data warehousing and OLAP.
61. Stumptner, R., Freudenthaler, B., & Krenn, M., (2012). *BIAccelerator–a Template-Based Approach for Rapid ETL Development.* Paper presented at the Foundations of Intelligent Systems: 20th International Symposium, ISMIS 2012, Macau, China, December 4–7, 2012. Proceedings 20.
62. Svilar, L., Castellano, J., Jukic, I., & Casamichana, D., (2018). Positional differences in elite basketball: Selecting appropriate training-load measures. *International Journal of Sports Physiology and Performance, 13*(7), 947–952.
63. Theodorou, V., Abelló, A., Lehner, W., & Thiele, M., (2016). Quality measures for ETL processes: From goals to implementation. *Concurrency and Computation: Practice and Experience, 28*(15), 3969–3993.
64. Thi, A. D. H., & Nguyen, B. T., (2008). A Semantic approach towards CWM-based ETL processes. *Proceedings of I-SEMANTICS, 8*, 58–66.
65. Trujillo, J., & Luján-Mora, S., (2003). *A UML Based Approach for Modeling ETL Processes in Data Warehouses.* Paper presented at the Conceptual Modeling-ER 2003: 22nd International Conference on Conceptual Modeling, Chicago, IL, USA, October 13–16, 2003. Proceedings 22.
66. Tziovara, V., Vassiliadis, P., & Simitsis, A., (2007). *Deciding the Physical Implementation of ETL Workflows.* Paper presented at the Proceedings of the ACM tenth international workshop on Data warehousing and OLAP.

67. Vassiliadis, P., Simitsis, A., & Baikousi, E., (2009). *A Taxonomy of ETL Activities.* Paper presented at the Proceedings of the ACM twelfth international workshop on data warehousing and OLAP.

68. Vyas, S., & Vaishnav, P., (2017). A comparative study of various ETL process and their testing techniques in data warehouse. *Journal of Statistics and Management Systems, 20*(4), 753–763.

69. Wang, H., & Ye, Z., (2010). *An ETL Services Framework Based on Metadata.* Paper presented at the 2010 2nd International Workshop on Intelligent Systems and Applications.

70. Wojciechowski, A., (2011). *E-ETL: Framework for Managing Evolving ETL Processes.* Paper presented at the Proceedings of the 4th workshop on Workshop for Ph. D. students in information & knowledge management.

71. Ying-lan, F., & Bing, H., (2009). *Design and Implementation of ETL Management Tool.* Paper presented at the 2009 Second International Symposium on Knowledge Acquisition and Modeling.

72. Yulianto, A. A., (2019). Extract transform load (ETL) process in distributed database academic data warehouse. *APTIKOM Journal on Computer Science and Information Technologies, 4*(2), 61–68.

CHAPTER

5

NoSQL Database

CONTENT

5.1. INTRODUCTION

There has been a growing trend towards adopting dispersed web-based applications in recent years. Services such as YouTube, Google, and Amazon are required to maintain and analyze vast quantities of data due to their widespread usage. It is evident that managing such data using single-node platforms is not feasible. Hence dispersed storage methods are imperative. The potential for expanding a system for managing relational databases (RDBMS) by adding nodes is frequently restricted or nonexistent (Han et al., 2011).

Numerous non-relational options for storage have appeared as a means of surpassing the constraints of scalability. Some instances of dispersed, non-relational records include Dynamo, which was created by Amazon, Google's The Big Table system, and Hadoop, which is utilized by Facebook. The expression "NoSQL" encompasses database platforms of this nature (Chandra, 2015) (Figure 5.1).

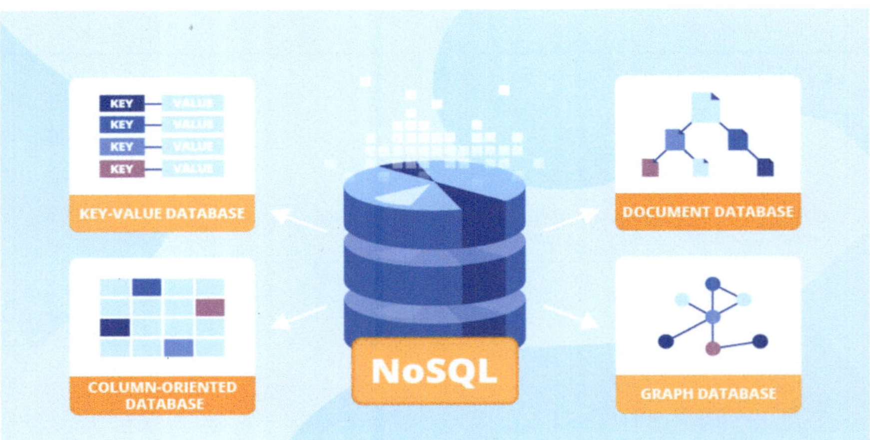

Figure 5.1. Illustration of components of NoSQL.

Source: Alex Williams, Creative Commons License.

This discourse delves into the fundamental tenets of systems that do not use SQL and their primary distinctions from relational database management systems (RDBMS). To commence, a brief overview of the relational framework will be presented. Subsequently, the following section presents the fundamental principles of NoSQL databases and provides a comprehensive survey of the NoSQL domain (Moniruzzaman & Hossain, 2013).

5.2. BASICS CONCEPTS

This text examines the fundamental attributes of conventional, SQL-centric systems to comprehend the needs that various NoSQL methodologies endeavor to fulfill.

5.2.1. Relational Databases

Edgar F. Codd, credited with inventing the relational structure, outlines in the fundamental constituents that establish a data model:

1. The term "**data structure**" refers to the various data types used to construct a model.
2. The **operators** refer to the set of available actions that can be performed to retrieve or control information contained in this particular structure.
3. **Integrity rules** refer to the fundamental regulations that establish coherent states for the data system.

Interactions, characteristics, tuples, and primary keys primarily define the relational information model's structure. In data management, it is common practice to represent relations in tables, where the attributes are organized as columns and the tuples are arranged as rows. The arrangement of attributes and tuples within the framework is not predetermined and, therefore, may be arbitrary (Strauch et al., 2011).

The utilization of LECT operations, which encompasses predictions and joins, is employed for data retrieval. Manipulative actions such as INSERT, UPDATE, and DELETE are also utilized for data manipulation.

The relational framework may be classified into two distinct categories of integrity rules. The enforcement of constraints, such as the requirement for primary keys to be unique, serves to maintain data integrity inside a given relation. Furthermore, independent referential integrity regulations exist among distinct relations (Okman et al., 2011).

5.2.2. ACID Properties

Transactions have been a crucial element in relational databases. Jim Gray established three fundamental attributes of a transaction in reference: atomic structure, consistency, and strength. Subsequently, Harder and Reuter condensed the characteristics and an additional isolation attribute into an abbreviation, ACID. While all four characteristics of ACID are widely recognized as crucial aspects of transactions within relational databases, uniformity is a particularly intriguing factor to consider when examining the capacity of a given system (Abramova et al., 2014) (Figure 5.2).

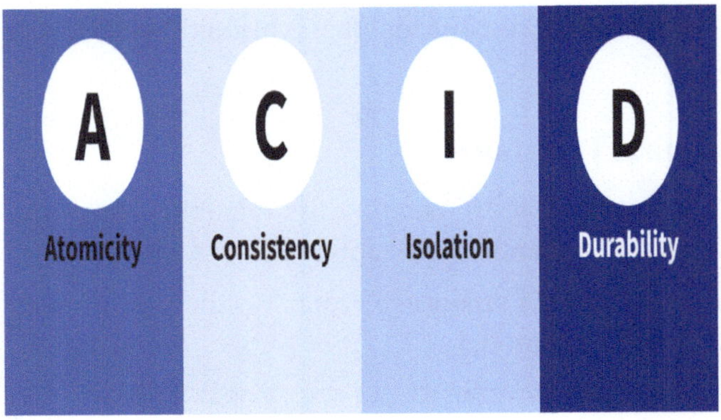

Figure 5.2. Schematic of the full form of ACID.

Source: Henrique Siebert Domareski, Creative Commons License.

5.2.3. Scalability

The capacity for a system to handle an increasing workload is commonly referred to because of its ability to scale. In essence, a system has the potential to be scaled in two distinct manners: vertically along with horizontally. Vertical scaling, also known as "scale up," involves the augmentation of a system's nodes' ability. Enhanced performance can be attained through the utilization of more robust hardware. On the contrary, horizontal scaling, also known as "scaling out," involves the addition of more nodes to a system (Stonebraker, 2010).

Did You Know?

NoSQL databases are designed to scale horizontally, allowing them to handle .massive amounts of data and high traffic loads efficiently

The vertical expansion of systems is often constrained by the accessibility and cost-effectiveness of superior gadgets or tends to exhibit inefficiency. On the contrary, the horizontal expansion of systems has been found to enhance efficiency while utilizing cost-effective hardware. However, the process of horizontal scaling is frequently a complex undertaking. The primary reason is the difficulty in achieving consistency guarantees, given by ACID characteristics, within distributed systems (Leavitt, 2010).

5.3. NOSQL CONCEPTS

The nomenclature "NoSQL" was initially coined by Carlo Strozzi to christen a database management system (DBMS) that he created. Although founded on a relational framework, the system in question deliberately refrained from utilizing SQL as its querying language. Currently, "NoSQL" is frequently understood as "Not only SQL" and pertains to primarily decentralized non-relational databases (Sicari et al., 2022).

Initially, an outline of the classifications of NoSQL databases is presented. Subsequently, the paper delves into the discussion of the reliability guarantees that are inherent to NoSQL databases.

The MapReduce paradigm is ultimately presented, serving as a robust structure for effectively handling vast quantities of data and as a crucial constituent of various NoSQL realizations (Tudorica & Bucur, 2011).

5.3.1. Types of NoSQL Systems

Edlich et al. (2016) have classified NoSQL systems into four categories, which they refer to as "Core-NoSQL" structures. These categories include Key-Value retail outlets, Wide column shops, Graph databases, and Document outlets.

This classification was presented in reference, page 6. Storage options associated with NoSQL, such as Object or XML files, are called "soft-NoSQL" systems. A comprehensive compilation of recognized NoSQL implementations, classified based on this nomenclature, is available in reference. The following section will provide a detailed explanation of only the "Core" courses (Nayak et al., 2013).

5.3.1.1. Key-Value Stores

Key-Value storage networks are a type of data structure that utilizes associated arrays comprised of distinct keys and corresponding values. To ensure unambiguous recognition of values, each key must be unique. Although keys are typically uncomplicated entities, values can encompass lists, sets, or even hashes, thereby facilitating the creation of more intricate structures for information (Figure 5.3).

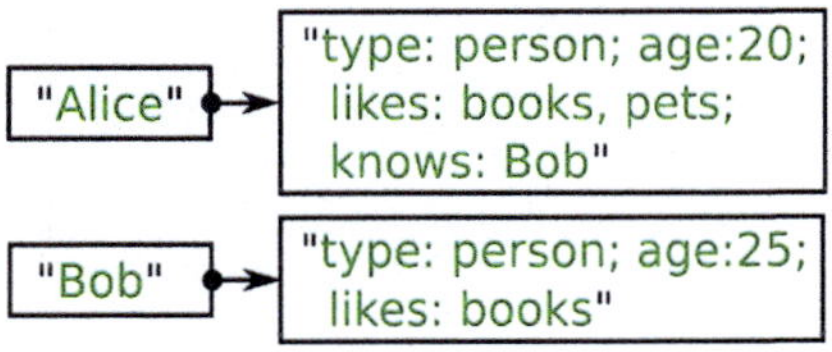

Figure 5.3. The present dataset is depicted within a key-value shop. Any information of the String data type is generally unable to be properly understood by the system that stores it.

Source: Science Direct, Creative Commons License.

The operations commonly available from Key-Value shops are analogous to those familiar to developers coming from Hash-Table executions (Reddy et al., 2022):

1. Add new pairs of keys and values.
2. Retrieve the value associated with a given key.
3. The removal of a key and its corresponding value.

The Key-Value store's straightforward model enables it to operate quickly and effectively. The trade-off for achieving a simplified structure is a concomitant limitation in the range of query options available. Amazon's Dynamo is a well-known exemplar of Key-Value stores. Dynamo provides an archive system that is both efficient along with extremely scalable. However, it is important to note that its search interface is significantly limited, as evidenced within the context above (Gudivada et al., 2014):

5.3.1.2. Wide Column Stores

The category of systems for storage being referred to is commonly known as Extensible Write down Stores. A wide section store may be conceptualized as a Key-Value shop, wherein the key is two-dimensional. A column key and a row's key are utilized for referencing a saved value. A date and time stamp may expand the key further, as shown in Google's Big Table. The number of potential variations for the key may vary according to the specific execution, often referred to as "keyspaces" or "domains." The keys in broad column storage possess numerous dimensions, leading to a configuration akin to a multifaceted, associated array (Haseeb & Pattun, 2017) (Figure 5.4).

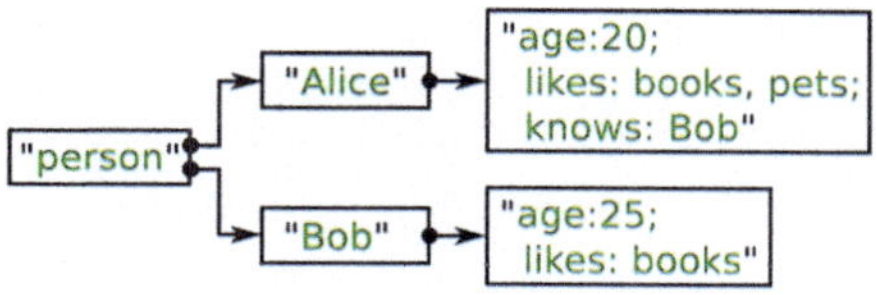

Figure 5.4. The information at hand is depicted in a columnar store format with a wide structure. The term "person" is employed as the column identifier, while the name of each individual is utilized as the row identifier. Like a key-value shop, the system must comprehend the stored value beyond its initial storage.

Science Direct, Creative Commons License.

5.3.1.3. Graph Databases

This system employs graphical representations to depict data, as its nomenclature suggests.

According to scholarly literature, databases containing graphs are suitable for representing data that exhibits a substantial and adaptable variety of connections, particularly when the knowledge of those connections holds equivalent significance to the portrayed data. Examples of such data may include social relationships or geographic information. Graph databases facilitate queries on the graph topology, such as the relationships among nodes or determining the shortest paths. Effective support with these queries can be achieved using established algorithmic methods and graph databases (Burckhardt, 2014) (Figure 5.5).

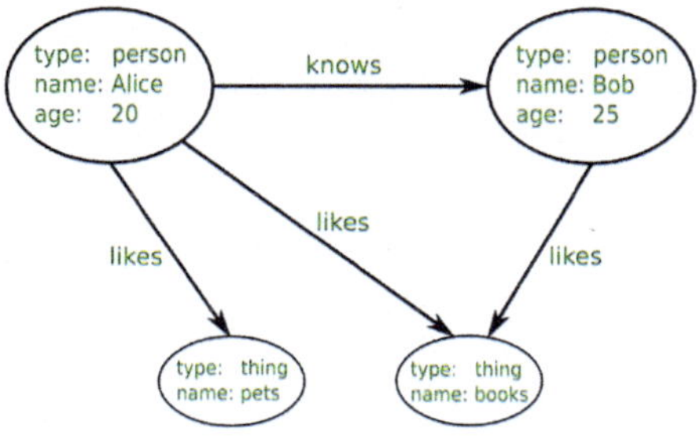

Figure 5.5. The graphical representation of data is exemplified in the following. Edges can serve as a relational data repository, whereas additional entities' characteristics must be stored within the vertices.

Source: Geeks for Geeks, Creative Commons License.

5.3.1.4. Document Stores

Information is retained in discrete pieces called documents within a paper-based database. The term "documents" within this context refers to any data structured in a certain manner. Examples of used formats are XML, BSON (see Section 1.4.1), and JSON. The specified data structure is often determined by the file that uses storage, although the data layout is flexible. As long as they adhere to the requirements of valid JSON files, it is conceivable to store disparate documents with different attribute sets in a Javascript-based records store. The data representation within a document's file, with JSON serving as the underlying data format. The data archiving system can understand the chosen data format. Instead of using a Key-information architecture, the system can enable exact read-and-write writing procedures on a document's features because it often lacks a deeper grasp of the stored information (Bailis & Ghodsi, 2013).

MongoDB, an example of the type of document storage system, is introduced in Section 1.4. CouchDB is a popular storage solution in this group that heavily depends upon the MapReduce architecture for data searching (Figure 5.6).

<table>
<tr><th colspan="2">people</th></tr>
<tr><td>{
name: "Alice",
age: 20,
likes: ["books", "pets"],
knows: "Bob"
}</td><td>{
name: "Bob",
age: 25,
likes: ["books"]
}</td></tr>
</table>

Figure 5.6. The document preserve is utilized for displaying data through examples. The data format utilized in this context is JSON. Direct requests on attributes such as "name" and "age" are now feasible due to the method's comprehension of the format.

Source: Geeks for Geeks, Creative Commons License.

5.3.2. Eventual Consistency

A significant contrast between relational databases and NoSQL systems pertains to the uniformity of the offered models. NoSQL databases must relax the ACID assurances provided by interpersonal transactions to facilitate straight scalability. Initially, this discourse elucidates three desirable attributes across distributed structures and comprehends:

- **Consistency:** The attainment of uniformity in a distributed system necessitates establishing a comprehensive order upon all operations executed across all nodes constituting the system, as posited by reference. For instance, this scenario would arise if all subsequent reading operations yield the created value after a triumphant compose operation, irrespective of the node within which the actions are carried out (Bermbach & Tai, 2011).

Tip

In NoSQL databases, eventual consistency means that after a certain period, all updates will propagate and be reflected consistently across the system. This approach .prioritizes availability and partition tolerance over immediate consistency

- **Availability:** The system is considered to fulfill the availability criterion when every operation carried out on a node within the system concludes with a response.
- **Partition Tolerance:** A divided system comprises a minimum of two distinct sets of nodes, wherein nodes belonging to a single set can interact. In contrast, messages exchanged among nodes belonging to various sets cannot reach their intended destination. An instance of a partitioned system could be observed when a node becomes inaccessible due to a network malfunction. A system can be deemed partition-accepting if it maintains both availability and consistency, despite the possibility of losing an arbitrary number of internal communications (Vogels, 2009).

Brewer proposed an idea in the year 2000, which posits that it is impossible for a web service to ensure reliability, accessibility, and tolerance for partitions simultaneously. After two years, the conjecture, commonly known as CAP Theorem and Brewer Theorem, was finally codified and established by Gilbert and Lynch. In practical terms, this implies that distributed databases must sacrifice one of these properties. To prevent partitions, it is necessary to ensure continuous accessibility of each node within a system to the other nodes. However, this assurance cannot be ascertained in large-scale distributed systems, as stated in reference. Therefore, regarding scalable systems, the choice ultimately comes to deciding between being available alongside consistent (Condie et al., 2010).

Pritchett suggests BASE as an alternative to ACID. The acronym BASE denotes the principles of being essentially available, possessing a soft state, and ultimately consistent. The primary emphasis of this approach is on ensuring the accessibility of a system, even if it means relaxing its assurances of uniformity. The BASE system's integrity and eventual uniformity allow clients to access data that is inconsistent or outdated during certain periods. However, it ensures that those time intervals will ultimately conclude. A system possessing BASE (Basically Available, Soft state, Eventually consistent) properties can overcome the constraints imposed by the CAP (Consistency, Availability, Partition tolerance) theorem, thereby providing significant potential for horizontal scaling. The chapter illustrates the irregular duration of instances for BASE structures using Amazon's S3 database as a case study. The authors have observed significant temporal fluctuations in inconsistency, ranging from a few milliseconds to moments (Dean & Ghemawat, 2008).

5.3.3. MapReduce

The MapReduce structure, which originates in Google, is a programming language and framework specifically developed to handle vast quantities of data through the application of user-defined logic.

The fundamental objective of the initial MapReduce architecture was to supply a conceptualization for handling data without encountering the complexities associated with scalabilities, such as parallelism load balancing, as stated in reference. The input for a MapReduce operation is a dictionary, a collection of data containing pairs of keys and values. The resulting data structure is once again a dictionary. The individual must implement two distinct functions, namely MAP and REDUCE and subsequently furnish these individuals with the structure (Hashem et al., 2016).

The MAP function is designed to accept a key and an amount as input parameters and generate an inventory of pairs of keys and values as an interim outcome.

The MapReduce paradigm executes the designated MAP procedure on each element of the input map. Given the independence of function calls, the structure parallelizes the processing per its requirements. The initial data is partitioned into multiple segments, individually processed by distinct MAP procedures (Karloff et al., 2010).

In the middle outcomes, all key-value combinations are aggregated based on the keys, resulting in an inventory of values for each key. The reduced operation is called upon for each key. The system is designed to accept a key and a corresponding list of values as its input. The REDUCE operation performs the computation of values and aggregates them into a conclusive result value, subsequently linked with the corresponding key. All instances of the REDUCE function can be executed concurrently. (Lee et al., 2012).

5.4. MONGODB

MongoDB has become an information database that is free to use and was created by 10gen, a software business that provides commercial support. The primary objective of MongoDB, as suggested by its nomenclature based on the term "humongous," is to manage vast quantities of data. Although MongoDB is coded in C++, it employs JavaScript as its primary query language.

The subsequent sections entail an analysis of the information model presented by MongoDB as the three fundamental constituents of data structures as previously defined. Most of the parts are derived from the present iteration of MongoDB's manual (Gyorodi et al., 2015).

5.4.1. Data Structure

MongoDB employs gathers as its fundamental units of data storage. A relationship (or table) is the interpersonal counterpart to a collection. Unlike relationships, an accumulation does not impose a rigid schema but can contain entirely disparate documents. Standardizing the structure for every record within a collection can facilitate streamlined and optimized data retrieval through querying (Boicea et al., 2012).

Documents might be likened to interpersonal tuples and regarded as arrays of associations or hashes. MongoDB employs a data storage and transmission mechanism that uses BSON (Binary JSON) structure for documents.

The BSON document primarily constitutes a binary manifestation of JSON paperwork, augmented with additional details to facilitate data parsing, such as duration prefixes. Like BSON documents, JSON items comprise attribute-value groups, wherein every value may be a basic type (such as string or integer), an intricate BSON object, or a set comprising either simple kinds or objects.

Moreover, BSON expands upon the JSON requirements by incorporating a few basic data types, such as those over time and date values. The comprehensive BSON requirement is available for reference in the following location (Chauhan, 2019).

5.4.2. Operators

5.4.2.1. Reading Data

The FIND functioning holds significant importance in retrieving information from a MongoDB collection. Without supplementary parameters, the function yields all records contained within a given being collected. The FIND function receives an argument as a standards object organized within BSON. MongoDB employs a filtering mechanism that selects documents based on prescribed criteria, exclusively presenting query outcomes that satisfy all designated attributes and corresponding values. Furthermore, MongoDB has defined specific conditioned operators that can be incorporated into the criteria. The operators are available for use within the database encompass a spectrum of functionalities, ranging from basic contrasts such as "lt" denoting "lesser than" to traditional expressing and intricate JavaScript functions that can be customized by the user and subsequently examined by the database system (Wei-ping et al., 2011).

An alternative method for retrieving data involves the utilization of the MAPREDUCE worker within MongoDB. The operator above requires a MAP and a REDUCE operation, provided in functions in JavaScript adhering to a stated format. As delineated in Section 1.3.3, the initial step involves the execution of the MAP function upon all the paperwork. Subsequently, the in-between outcomes are subjected to processing via the REDUCE function. The outcomes yielded by the REDUCE operation signify the outcome of the query. Listing 1.2 illustrates a specific instance (Lee et al., 2004).

```
db.people.find({name: ‘Alice’})
// {_id: 1234, name: ‘Alice,’ age: 20, likes: [‘ books,’ ‘pets’], knows: [‘Bob’]}
db.people.find({age: 25})
// {_id: 5678, name: ‘Bob,’ age: 25, likes: [‘books
‘]}
db.people.find({age: {$lt: 30}})
// [{_id: 1234, name: ‘Alice,.’ }, {_id: 5678, name: ‘Bob,.’ }]
db.people.find({knows: {$exists: true}})
// {_id: 1234, name: ‘Alice,.’ }
db.people.find({likes: {$size: 1}})
// {_id: 5678, name: ‘Bob,.’ }
```

Listing 1.1 The following are sample queries utilizing the FIND function. It is assumed that the collection of individuals being queried, referred to as "people," comprises two distinct entries, namely "Alice" and "Bob."

5.4.2.2. Manipulating Data

Incorporating new documents into a collection is achieved through INSERT operations. The operational procedure receives the document for insertion in BSON format. Before storing the provided paperwork, MongoDB appends an extra attribute labeled "id" with a distinct value across the entire collection. Those above may be regarded as the main key, serving the purpose of distinguishing the document. The UPDATE functioning can be employed to modify pre-existing objects within an assortment. The function requires a criteria object and an object representing the implemented modifications. If the latter object denotes a standard document, the initial document that satisfies the specified criteria undergoes a complete replacement. MongoDB offers supplementary alteration syntax to modify particular characteristics of matched paperwork exclusively. Listing 1.3 presents several examples (Gao & Qiu, 2010). The act of removing documents can be achieved by utilizing the REMOVE operations. The requirements object may target the documents intended to be removed.

```
map = function() {
for (i in this.likes) {
emit(this.likes[i], {count: 1});
}
};
reduce = function(key, values) {
var total = 0;
for (i in values) {
total += values[i].count
}
return {count: total}
};
db.people.mapReduce(map, reduce)
// [{ books: {count: 3} },
// { pets: {count: 4} }]
```

5.5. INTEGRITY RULES

In contrast to relational databases, which provide diverse honesty regulations, MongoDB provides options for enforcing document distinctiveness. As previously stated, every document possesses an "id" attribute. Distinct characteristics, along with amalgamations of characteristics, have the potential to be designated as exclusive indices as well, as outlined in Section 1.4.4. The implementation of unique indexes guarantees that the values assigned to the designated attributes remain distinct across the entirety of a given collection (Chang et al., 2015).

MongoDB lacks inherent assistance with record joining and provides no guidelines for maintaining the integrity of referential relationships.

5.6. INDEXING

Database searches are a crucial component in enhancing the efficiency of examining queries. By standard, the primary essential section in a relational database is usually indexed. The statement above holds valid for MongoDB paperwork's "id" link.

```
db.people.insert({name: 'Alice,' age: 20})
db.people.update({name: 'Alice'}, {name: 'Eve'})
db.people.find({name: 'Eve'})
// {_id: 123, name: 'Eve'}
db.people.insert({name: 'Alice,' age: 20})
db.people.update({name: 'Alice'}, {$set: {name: 'Eve
'}})
db.people.find({name: 'Eve'})
// {_id: 456, name: 'Eve,' age: 20}
db.people.insert({name: 'Alice,' age: 20})
db.people.update({name: 'Alice'}, {$inc: {age: 5}})
db. people.find({name: 'Alice'})
// {_id: 789, name: 'Alice,' age: 25}
```

Listing 1.3: This chapter provides examples of manipulating documents within MongoDB utilizing the INSERT and UPDATE operations.

MongoDB provides the capability to create numerous indexes for a collection, enabling efficient data retrieval. It is possible to index every single attribute of paperwork and a variety of attributes. This pertains to characteristics present in sub-documents, including attributes that are nested within other attributes) (Hikitih et al., 2021).

As previously mentioned, the arrangement of an item is not predetermined by the assembly. Because indices are generated on a per-collection basis, it is possible for certain documents throughout a collection to lack a particular attribute even though that attribute has an index. MongoDB treats such documents as having the attribute with a value of *null*. For collections containing strongly varying documents, this causes a significant overhead. The presence of highly diverse paperwork within a collection results in a substantial increase in overhead. Hence, MongoDB presents an alternative indexing mechanism, namely sparse indexing. In cases where an index has become sparse, it disregards any documents within a collection that lack the indexed characteristic. This feature enables improved performance and decreased storage overhead by eliminating the need for creating and upholding "empty" indexes (Indrawan, 2012).

According to relational database systems, it is well-established that while indexes enhance the efficiency of reading operations, they also increase the cost of writing operations. Each time a document undergoes creation, modification, or deletion, it is necessary to update the corresponding indexes. Therefore, particularly in scenarios with a high volume of writing, generating an excessive number of indexes could potentially have a notable adverse impact on the system's overall efficiency (Guo & Onstein, 2020).

5.6.1. Scalability

MongoDB cannot manage inter-collection references, necessitating that every client-initiated read or write operation be restricted to a single collection. Vertical expansion of applications that involve numerous smaller numbers may be accomplished by distributing the collections across multiple machines.

Remember

Indexing plays a crucial role in achieving scalability in NoSQL databases by optimizing query performance and enabling efficient data retrieval based on indexed .fields

Nevertheless, the process of vertical scaling becomes increasingly intricate when managing vast quantities of documents that are stored within a singular collection. The MongoDB platform provides a sharding mechanism that enables the distribution of collections across multiple nodes in a manner forthcoming to the application. To execute this task, it is necessary to specify the number of fragments to be utilized and a designated shard key (Gessert et al., 2017).

For a collection to be shared, its papers must possess a shared key characteristic. Utilizing a fragmented key facilitates the partitioning of a collection through numerous segments. A chunk refers to a distinct portion of a given collection that contains a set of documents. The set of paperwork is defined by the values that comprise the fragmented key falling throughout a specific range (Mahmood et al., 2015).

Since the attribute is designated as the shard key, a collection comprising n chunks $(c_i)_{i=1.n}$ will include all documents alongside principles v for attribute a within each chunk ci, where $min_i \leq v < max_i$. Determining the fragment key varies for every chunk. Specifically, those over mini and maxi is a system-driven process to ensure equitable distribution of every record across accessible shards. Upon exceeding a predetermined threshold, the network undergoes a process of chunk division and redistribution to maintain equitable load distribution across all shards (Diogo et al., 2019).

The efficacy of reading and writing on a sharded gathering is heavily contingent upon selecting an appropriate shard key. When an attribute is utilized as a shard key and has a uniform value across most paperwork, it results in the corresponding chunk containing those files being impervious to splitting. As per reference, it is not possible to evenly distribute the assortment uniformly among all shards. If the individuals in a collection are predominantly compared to the same country, utilizing a characteristic such as "home country" as a shard key would not be optimal. In this scenario, MongoDB could not establish a uniformly distributed splitting (Xiang et al., 2010).

5.6.2. Consistency

MongoDB provides excellent consistency on individual entities when there is just a single server managing all the reading and writing demands across every shard. This implies that the atomicity property applies to all reading and writing actions performed on individual papers and that any FIND activities involving only a single record are ensured to accurately reflect the most current state of the database (Abramova & Bernardino, 2013).

Furthermore, it is possible to set a choice in MongoDB's FIND activities that permit reading from secondary sources. If replication to occur is employed, a copied node, also known as an enslaved person, has the potential to respond to the inquiry. Thus, the accessibility of the information within it can be enhanced. However, this trade-off entails a compromise in uniformity, as stipulated within the CAP theorem. Due to the possibility of the copied node retaining obsolete data about the entity in question, the only form of uniformity that can be ensured is future consistency (Manoj, 2014).

Activity 5.1

As a software developer, you are tasked with designing a social media platform that requires handling a massive amount of user-generated content. Outline the key reasons for considering a NoSQL database over a traditional relational database in this scenario. Discuss the concept of scalability and eventual consistency in NoSQL databases and explain how they can address the challenges of managing .and processing large volumes of dynamic user data

5.7. SUMMARY

The relational database structure is a highly adaptable storage solution that can be applied to various issues, owing to its intricate array of employees and integrity regulations. Furthermore, relational transactions, which possess ACID properties, provide robust assurances regarding consistency. In contrast, NoSQL storage options generally strive to offer streamlined yet highly effective resolutions for particular issues. The primary objective of numerous NoSQL databases is to offer exceptional scalability. To prioritize the accessible nature of a system, it may be necessary to ease the strong uniformity guarantees typically associated with relational systems. Consequently, the concept of eventually remaining consistent is derived.

MongoDB, a document-oriented database, has been extensively discussed among the vast NoSQL systems. In contrast to relational systems that operate within rigid data structures, document stores facilitate the storage of diverse objects cohesively, provided they conform to a uniform data format. Document archives cannot typically accommodate intricate relationships among stored entities. A potential solution to this issue could involve accepting denormalized data, which entails the storage of interconnected data redundantly. In contrast to certain NoSQL systems, such as CouchDB, that impose restrictions on data searching and necessitate using MapReduce for handling information, MongoDB provides a versatile

API that employs a criteria syntax similar to JSON. However, it is worth noting that the flexibility offered by relational systems through the SQL programming language has yet to be surpassed.

No single NoSQL system can serve as a universal storage solution or supplant a system with a relational the selection of an appropriate system is necessary to address the given task effectively. When expedient data retrieval is required and querying through a solitary key is satisfactory, KeyValue stores prove to be a commendable option. Wide column stores enable the creation of more intricate key structures. A database based on graphs is most suitable for storing objects that are highly interconnected yet simple and for queries that pertain to the structure of relationships. On the contrary, querying interactions in document-oriented structures may be more complex. Document stores enable the storage of complexly structured objects in a scalable manner and provide effective mechanisms for querying such structures.

REVIEW QUESTIONS

1. What are the fundamental differences between NoSQL databases and relational databases?
2. Explain the ACID properties and their role in relational databases.
3. Discuss the concept of scalability in the context of NoSQL databases.
4. What are the different types of NoSQL systems, and how do they differ in their data models?
5. Define eventual consistency and its significance in distributed systems.
6. How does MapReduce contribute to data processing in NoSQL databases?
7. Explore the data structure and operators in MongoDB.
8. Explain the importance of integrity rules in NoSQL databases.
9. Discuss the role of indexing in achieving scalability in NoSQL databases.

What are the challenges in maintaining consistency in distributed NoSQL databases?

REFERENCES

1. Abramova, V., & Bernardino, J., (2013). NoSQL databases: MongoDB vs Cassandra. In: *Proceedings of the International C* Conference on Computer Science and Software Engineering, 1*(1), (pp. 14–22).
2. Abramova, V., Bernardino, J., & Furtado, P., (2014). Which NoSQL database? a performance overview. *Open Journal of Databases (OJDB)*, *1*(2), 17–24.
3. Bailis, P., & Ghodsi, A., (2013). Eventual consistency today: Limitations, extensions, and beyond. *Communications of the ACM*, *56*(5), 55–63.
4. Balegas, V., Duarte, S., Ferreira, C., Rodrigues, R., Preguiça, N., Najafzadeh, M., & Shapiro, M., (2015). Putting consistency back into eventual consistency. In: *Proceedings of the Tenth European Conference on Computer Systems* (pp. 1–16).
5. Bermbach, D., & Tai, S., (2011). Eventual consistency: How soon is eventual? An evaluation of Amazon S3's consistency behavior. In: *Proceedings of the 6th Workshop on Middleware for Service Oriented Computing* (pp. 1–6).
6. Boicea, A., Radulescu, F., & Agapin, L. I., (2012). MongoDB vs oracle–database comparison. In: *2012 Third International Conference on Emerging Intelligent Data and Web Technologies* (Vol. 1, No. 1, pp. 330–335).
7. Burckhardt, S., (2014). Principles of eventual consistency. *Foundations and Trends® in Programming Languages*, *1*(1, 2), 1–150.
8. Burdakov, A., Grigorev, U., Ploutenko, A., & Ttsviashchenko, E., (2016). Estimation models for NoSQL database consistency characteristics. In: *2016 24th Euromicro International Conference on Parallel, Distributed, and Network-Based Processing (PDP)* (Vol. 1, No. 1, pp. 35–42).
9. Chandra, D. G., (2015). BASE analysis of NoSQL database. *Future Generation Computer Systems*, *52*, 13–21.
10. Chang, B. R., Tsai, H. F., Wang, Y. A., & Chen, C. Y., (2015). Realization of secondary indexing to NoSQL database with intelligent adaptation. In: *2015 Conference on Technologies and Applications of Artificial Intelligence (TAAI)* (pp. 449–452). IEEE.
11. Chauhan, A., (2019). A review on various aspects of MongoDB databases. *International Journal of Engineering Research & Technology (IJERT)*, *8*(5), 90–92.

12. Condie, T., Conway, N., Alvaro, P., Hellerstein, J. M., Elmeleegy, K., & Sears, R., (2010). MapReduce online. In: *NSDI* (Vol. 10, No. 4, p. 20).
13. Dean, J., & Ghemawat, S., (2008). MapReduce: Simplified data processing on large clusters. *Communications of the ACM*, *51*(1), 107–113.
14. Diogo, M., Cabral, B., & Bernardino, J., (2019). Consistency models of NoSQL databases. *Future Internet*, *11*(2), 43.
15. Gao, X., & Qiu, J., (2010). *Scalable Inverted Indexing on NoSQL Table Storage* (Vol. 1, No. 1, pp. 1–20). A Class.
16. George, S., (2013). NoSQL–NOT ONLY SQL. *International Journal of Enterprise Computing and Business Systems*, *2*(2), 1–20.
17. Gessert, F., Wingerath, W., Friedrich, S., & Ritter, N., (2017). NoSQL database systems: A survey and decision guidance. *Computer Science-Research and Development*, *32*, 353–365.
18. Gudivada, V. N., Rao, D., & Raghavan, V. V., (2014). NoSQL systems for big data management. In: *2014 IEEE World Congress on Services* (Vol. 1, No. 1, pp. 190–197).
19. Guo, D., & Onstein, E., (2020). State-of-the-art geospatial information processing in NoSQL databases. *ISPRS International Journal of Geo-Information*, *9*(5), 331.
20. Győrödi, C., Győrödi, R., Pecherle, G., & Olah, A., (2015). A comparative study: MongoDB vs. MySQL. In: *2015 13th International Conference on Engineering of Modern Electric Systems (EMES)* (Vol. 1, No. 1, pp. 1–6).
21. Han, J., Haihong, E., Le, G., & Du, J., (2011). Survey on NoSQL database. In: *2011 6th International Conference on Pervasive Computing and Applications* (pp. 363–366). IEEE.
22. Haseeb, A., & Pattun, G., (2017). A review on NoSQL: Applications and challenges. *International Journal of Advanced Research in Computer Science*, *8*(1), 1–20.
23. Hashem, I. A. T., Anuar, N. B., Gani, A., Yaqoob, I., Xia, F., & Khan, S. U., (2016). MapReduce: Review and open challenges. *Scientometrics*, *109*, 389–422.
24. Indrawan-Santiago, M., (2012). Database research: Are we at a crossroad? Reflection on NoSQL. In: *2012 15th International*

Conference on Network-Based Information Systems (pp. 45–51). IEEE.
25. Karloff, H., Suri, S., & Vassilvitskii, S., (2010). A model of computation for MapReduce. In: *Proceedings of the Twenty-First Annual ACM-SIAM Symposium on Discrete Algorithms* (pp. 938–948). Society for Industrial and Applied Mathematics.
26. Leavitt, N., (2010). Will NoSQL databases live up to their promise? *Computer*, *43*(2), 12–14.
27. Lee, K. H., Lee, Y. J., Choi, H., Chung, Y. D., & Moon, B., (2012). Parallel data processing with MapReduce: A survey. *ACM SIGMOD Record*, *40*(4), 11–20.
28. Lee, Y. W., Pipino, L., Strong, D. M., & Wang, R. Y., (2004). Process-embedded data integrity. *Journal of Database Management (JDM)*, *15*(1), 87–103.
29. Mahmood, K., Risch, T., & Zhu, M., (2015). Utilizing a NoSQL data store for scalable log analysis. In: *Proceedings of the 19th International Database Engineering & Applications Symposium* (Vol. 1, No. 1, pp. 49–55).
30. Manoj, V., (2014). Comparative study of NoSQL document, column store databases and evaluation of Cassandra. *International Journal of Database Management Systems*, *6*(4), 11.
31. Moniruzzaman, A. B. M., & Hossain, S. A., (2013). *NoSQL Database: New Era of Databases for Big Data Analytics-Classification, Characteristics and Comparison*. arXiv preprint arXiv:1307.0191.
32. Nayak, A., Poriya, A., & Poojary, D., (2013). Type of NOSQL databases and its comparison with relational databases. *International Journal of Applied Information Systems*, *5*(4), 16–19.
33. Okman, L., Gal-Oz, N., Gonen, Y., Gudes, E., & Abramov, J., (2011). Security issues in NoSQL databases. In: *2011IEEE 10th International Conference on Trust, Security and Privacy in Computing and Communications* (Vol. 1, No. 1, pp. 541–547).
34. Reddy, H. B. S., Reddy, R. R. S., Jonnalagadda, R., Singh, P., & Gogineni, A., (2022). Analysis of the unexplored security issues common to all types of NoSQL databases. *Asian Journal of Research in Computer Science*, *14*(1), 1–12.
35. Schindler, J., (2012). I/O characteristics of NoSQL databases. *Proceedings of the VLDB Endowment*, *5*(12), 2020, 2021.

36. Schindler, J., (2013). Profiling and analyzing the I/O performance of NoSQL DBs. In: *Proceedings of the ACM SIGMETRICS/International Conference on Measurement and Modeling of Computer Systems* (Vol. 1, No. 1, pp. 389–390).
37. Sicari, S., Rizzardi, A., & Coen-Porisini, A., (2022). Security & privacy issues and challenges in NoSQL databases. *Computer Networks, 1*(1), 108828.
38. Stonebraker, M., (2010). SQL databases v. NoSQL databases. *Communications of the ACM, 53*(4), 10–11.
39. Strauch, C., Sites, U. L. S., & Kriha, W., (2011). NoSQL databases. *Lecture Notes, Stuttgart Media University, 20*(24), 79.
40. Swaminathan, S. N., & Elmasri, R., (2016). Quantitative analysis of scalable NoSQL databases. In: *2016 IEEE International Congress on Big Data (BigData Congress)* (Vol. 1, No. 1, pp. 323–326).
41. Tudorica, B. G., & Bucur, C., (2011). A comparison between several NoSQL databases with comments and notes. In: *2011 RoEduNet International Conference 10th Edition: Networking in Education and Research* (Vol. 1, No. 1, pp. 1–5).
42. Vogels, W., (2009). Eventually consistent. *Communications of the ACM, 52*(1), 40–44.
43. Wei-Ping, Z., Ming-Xin, L. I., & Huan, C., (2011). Using MongoDB to implement textbook management system instead of MySQL. In: *2011 IEEE 3rd International Conference on Communication Software and Networks* (Vol. 1, No. 1, pp. 303–305).
44. Xiang, P., Hou, R., & Zhou, Z., (2010). Cache and consistency in NOSQL. In: *2010 3rd International Conference on Computer Science and Information Technology* (Vol. 6, pp. 117–120).
45. Yu, Q., & Zhou, J., (2022). Consistent and efficient batch operations for NoSQL databases with hybrid timestamp. In: *Network and Parallel Computing: 19th IFIP WG 10.3 International Conference, NPC 2022, Jinan, China, September 24–25, 2022, Proceedings* (Vol. 1, No. 1, pp. 340–351). Cham: Springer Nature Switzerland.
46. Nikitin, V., Krilov, Y., Kornaga, Y., & Anikin, V. (2021). Combined indexing method in nosql databases. Adaptive automatic control systems, 1(38), 3-9.

CHAPTER 6

Cloud Databases and Big Data

CONTENT

6.1. INTRODUCTION

Big Data refers to a methodology for analyzing data that is made possible by advancements in technology and architecture. This methodology is characterized by the ability to capture, store, and analyze data at high velocities. According to scholarly sources, the range of data sources has expanded beyond the conventional corporate database to encompass diverse forms of information such as email, output from mobile devices, data generated by sensors, and output from social media. Contemporary data encompasses not only structured records stored in databases, but also unstructured data that lacks standardized formatting (Awiti et al., 2020).

The processing and analysis of Big Data necessitates a significant allocation of storage capacity. Although the cost of storage has been decreasing, small and medium-sized enterprises may still encounter financial challenges in utilizing big data due to the required resources. A commonly employed infrastructure for the storage and analysis of big data involves the utilization of crowded network-attached storage (NAS). The establishment of a crowded NAS infrastructure necessitates the arrangement of multiple NAS "pods," each of which consists of multiple storage gadgets that are linked to a NAS device. The NAS appliances are subsequently interconnected to facilitate extensive data sharing along with searching (Albrecht & Naumann, 2008) (Figure 6.1).

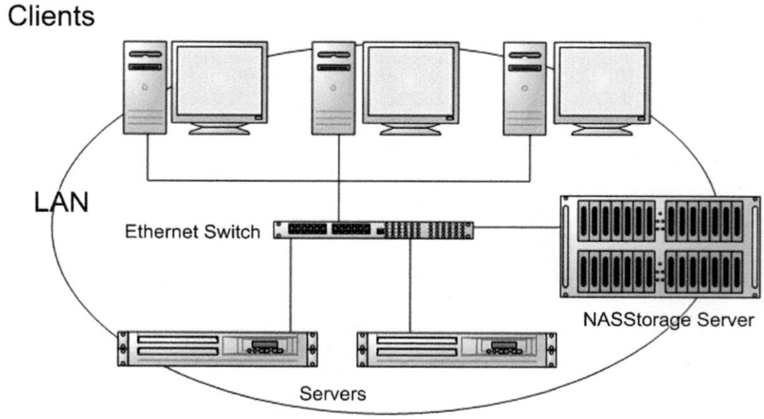

Figure 6.1. Illustration of network-attached storage (NAS).

Source: Tech-FAQ, Creative Commons License.

The approach of service-oriented computation called cloud computing achieved remarkable success and brought about a significant transformation in the abstraction and utilization of computing infrastructure. The three most widely recognized cloud paradigms can be Infrastructure to be a Service, Platforms to be a Service, as well as Software be a Service. The aforementioned concept can be further expanded to encompass a Database in the form of a Service (DBaaS) or Storage space as a Service (STaaS) (Figure 6.2).

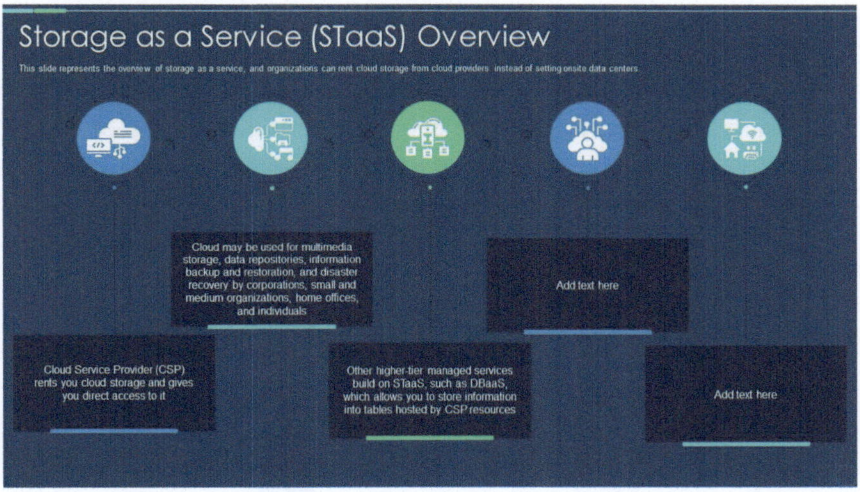

Figure 6.2. Illustration of storage space as a service (STaaS).

Source: Slide team, Creative Commons License.

Cloud computing has become a widely accepted paradigm for deploying innovative applications that had been previously not affordable in traditional business infrastructure settings. This is due to several enabling features like flexibility, pay-per-use, minimal upfront investments, minimal time to markets, as well as the transfer of potential hazards. Database management systems that can be scaled to accommodate update-intensive apps and decision-support tools are crucial components of cloud infrastructure. For over thirty years, the field of database analysis has aimed to achieve scalable as well as dispersed data management. Considerable scholarly attention has been devoted to the development of scalable systems that can accommodate the two: update-intensive tasks and ad-hoc analysis tasks. The preliminary designs encompass the utilization of databases that are distributed for tasks that require frequent updates, as well as parallel database structures for workloads that involve analysis. The development of parallel databases progressed from prototype systems to extensive commercial systems. However, distributed database platforms did not achieve significant success and weren't commercialized. Instead, diverse ad-hoc methods were employed to facilitate scaling (Bao et al., 2007).

The emergence of Key-Value outlets can be attributed to alterations within the data accessibility patterns of apps and the necessity to expand to numerous commodity machines. This has resulted in the development of a new category of systems that are being extensively embraced by diverse enterprises. The MapReduce framework and which is open-source execution Hadoop have been extensively adopted in businesses as well as universities for data analysis purposes. Proposals have been put forth to enhance the usability as well as efficiency of Hadoop-based mechanisms (Galici et al., 2020) (Figure 6.3).

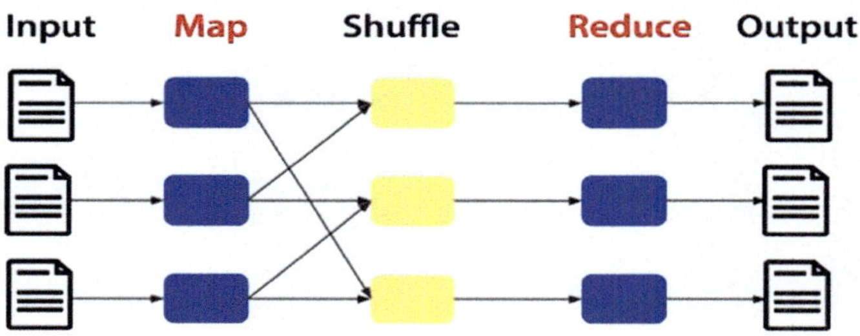

Figure 6.3. Schematic of MapReduce framework.

Source: EduPristine, Creative Commons License.

Conversely, data centers have been employed for the purpose of managing vast quantities of data. The existing warehouses and their associated solutions are inadequate in terms of providing timely responses to cope with the increasing volumes of data. There are two approaches to data processing: batch analytics on large volumes or real-time transactions on small data sets. Given the updated specifications, it is imperative to guarantee a prompt and efficient response, either in real-time or nearly real-time, for vast quantities of data. The four dimensions of big data, namely the amount, speed, variety, as well as veracity, pose significant challenges to conventional data warehousing systems in terms of data management along with analytics. The term "big data" refers to data sets that surpass the processing capabilities of traditional database systems. The aforementioned statement suggests that the quantity of data is excessive, and/or the data values undergo rapid changes, and/or it fails to conform to the established principles of traditional database management systems, such as consistency. Professionals in the field of information and technology management must possess novel competencies that enable them to proficiently model along with prepare data for assessment. Additionally, they must possess a deep understanding of the problem at hand to effectively execute analytics. Due to the vast size and dynamic nature of data, a substantial amount of memory and central processing unit resources are required. These resources can be obtained through the utilization of distributed processors as well as storage space in cloud environments (Lee & Park, 2021).

The utilization of cloud computing over data storage seems a feasible alternative for small and medium enterprises that are contemplating the implementation of Big Data analytical methodologies. Cloud computing refers to the provision of computing resources by an external entity, which can be accessed on-demand through a network (Figure 6.4).

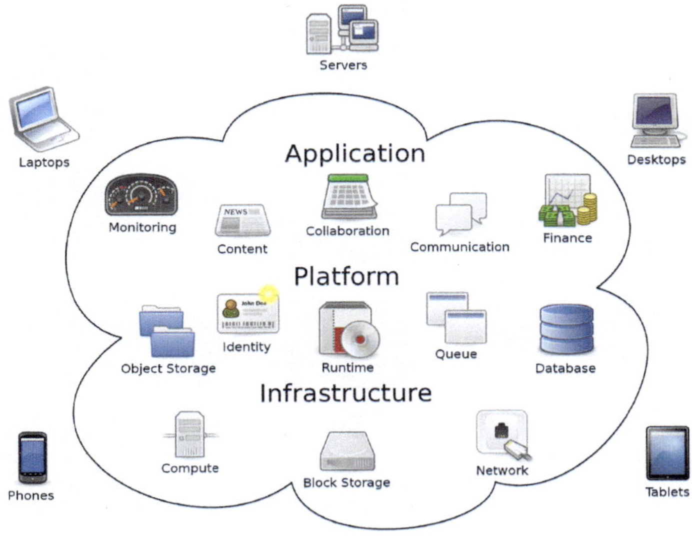

Figure 6.4. Image of components of cloud computing.

Source: Sam Johston, Creative Commons License.

This approach typically requires minimal management effort from the business. Cloud computing encompasses various architectures as well as deployment models that can be integrated with diverse technologies along with design methodologies. Small and medium-sized business proprietors who lack the financial resources to implement clustered NAS technologies may explore various cloud computing algorithms to fulfill their big data requirements. It is imperative for small as well as medium-sized enterprise proprietors to contemplate the appropriate cloud computing solution to sustain their competitiveness and profitability (Rodriguez-Muro et al., 2012).

6.2. BIG DATA

The term "Big Data" is utilized to describe the inadequacy of conventional data frameworks in effectively managing contemporary data collections. The need for a novel architecture to attain efficiencies is driven by certain characteristics. These include the amount and diversity of data available from diverse domains or forms, which pertain to the data itself set at rest. Additionally, the data-in-motion features such as velocity, the degree of flow, as well as variability (primarily denoting a change within velocity) also play a crucial role. Diverse attributes give rise to distinct architectures or varying sequences of data lifecycle processes that are necessary to attain the required efficiencies. Several alternative terms (frequently commencing with the word 'V') are utilized, however, a subset of these pertain to analytics rather than big data structures (Silva et al., 2016) (Figure 6.5).

Figure 6.5. Illustration of big data.

Source: Surya Gutta, Creative Commons License.

The emergence of the big data framework is characterized by the need to prioritize managing all kinds of data as a crucial factor in the development of system architecture, particularly when dealing with large volumes of data that are either at rest or constantly in motion. The big data framework entails a fundamental transformation in data system structures, moving away from monolithic structures that rely on vertical scaling (i.e., more powerful processors or disks) and towards horizontally scaled structures that incorporate a loosely integrated collection of resources. Approximately two decades ago, a transition took place within the modeling community whereby scientific simulations started utilizing massively parallel processing (MPP) networks.

By employing diverse arrangements of segregating the code as well as data among autonomous processors, researchers in computational science succeeded in significantly enhancing their simulation capacities (Uddin & Gupta, 2014).

The aforementioned has resulted in several intricacies in domains such as communication of messages, transfer of data, delay in uniformity across resources, distribution of workload, and operational inadequacies while anticipating the completion of tasks by other resources. Similarly, the big data framework embodies a comparable transformation, employing distinct mechanisms to disseminate code and data among loosely-coupled resources to facilitate the requisite scaling in data management that corresponds to the scaling within the data.

The primary objective of the storage and retrieval of substantial volumes of data has been to conduct a study that generates supplementary insights into the data. Historically, data analysis usually took place on a randomly selected subset of the dataset (Bakshi, 2012).

6.2.1. Big Data Engineering

Big Data Technology refers to the utilization of storage as well as data manipulation technology that make use of a set of horizontally connected resources in order to attain performance scalability which is almost linear.

The increasing significance of data types that aren't amenable to effective handling within a conventional relational model has propelled the development of novel engineering techniques within the data layers. The requirement for expandable accessibility in both organized and unorganized data has resulted in the development of software centered around name-value/key-value pairs or just columnar (big table), document-centric, along with graph (that includes triple-store) models (Chacko et al., 2015).

6.2.2. Non-Relational Model

Non-Relational Simulations pertain to logic-based models, including records, graphs, key-value, as well as additional models, which are utilized to facilitate more effective storage and retrieval of non-tabular databases.

6.2.3. NoSQL

NoSQL, also known as "no SQL" or just "not only SQL," pertains to data storage and interfaces that do not adhere to rigid relational methodologies (Davoudian et al., 2018) (Figure 6.6).

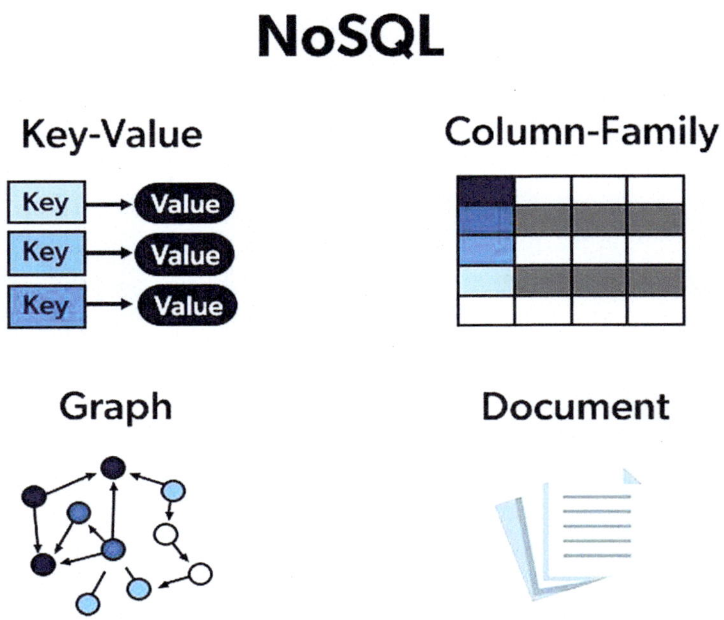

Figure 6.6. Image of elements of NOSQL.

Source: Hiteshreddy, Creative Commons License.

6.2.4. Big Data Models

The term "Big Data Models" pertains to conceptual data models, including both relational as well as non-relational models, as well as the processing along with computation models such as batch, streaming, as well as transactional models. These models are utilized for the purpose of storing and manipulating data within horizontally developed resources (Sharma et al., 2014).

Did You Know?

Big Data models encompass a wide range of data types, with unstructured data accounting for approximately 80% of the total data generated globally

6.2.5. Schema-on-Read

Big data that follows a schema-on-read approach is frequently stored within an unprocessed state according to the method of production. The schema required for arranging and frequently refining the data is identified and modified during the querying process. The structuring of data for supporting particular algorithms or just processing structures is crucial for the effective operation of numerous analytics (Janković et al., 2018).

6.2.6. Big Data Analytics

The field of Big Data Analysis has undergone swift development with respect to its functionality as well as the fundamental programming paradigm. Analytical functions facilitate the integration of parallel-derived results from distributed segments of one or multiple data sources (Tsai et al., 2015).

6.2.7. Big Data Paradigm

The big data framework is characterized by the deployment of data systems that are horizontally distributed throughout independent resources, which enables the scalability required for the effective processing of large-scale data sets.

The emergence of the Big Data Framework has enabled the performance of analytical operations on the complete data set or possibly on an ongoing flow of data within real-time. The process of analysis has the potential to incorporate diverse data sources originating from various entities. An illustrative inquiry pertains to the correlation that exists among insect-borne ailments, temperature, precipitation, as well as variations in foliage. In order to address this inquiry, a comprehensive analysis would necessitate the integration of data pertaining to disease incidence along with geographic distribution, meteorological data, as well as aerial imagery (Sun et al., 2018). It is anticipated that there will be further advancements in the techniques employed to attain optimal scalability across various resources. However, the current shift towards the utilization of several horizontally-coupled facilities for processing or data systems, commonly referred to as "crowdsourcing," can be likened to a paradigm shift. This shift is expected to be a singular event unless another paradigm shift arises in the future. This is similar to the previous paradigm shift observed in the simulated community.

The technical improvements associated with the Big Data framework have additional implications. The modifications are not solely confined to rational data retention but also encompass the concurrent dissemination of data as well as code within the tangible file system, along with direct interrogations directed toward this retention (Patgiri & Ahmed, 2016).

The alteration in cognitive processes results in modifications to the conventional data life cycle. The entire data lifecycle can be delineated into four distinct stages, namely, collection, preparation, evaluation, as well as action. Big data utilization scenarios can be classified based on the attributes of data sets in a stationary or mobile state and with regard to the duration of the complete data lifecycle. The features of a collection of data have varying impacts on the processes involved in the data lifecycle. One such impact is observed in the stage of the lifecycle where the data is stored in a persistent manner. In the conventional relational paradigm, the data undergoes preparation procedures such as extract-transform-load as well as a cleansing before being stored. In scenarios where there is a need for rapid data processing, the information is first subjected to preparation and analysis to enable timely alerting. Subsequently, the data or its aggregates are stored persistently. In a scenario involving voluminous data, it is common practice to store the data in its original form, prior to undergoing any preparatory procedures aimed at refining and structuring the data. The ongoing existence of data within its raw form implies that the application of an arrangement or representation of the data occurs exclusively during data retrieval, which is commonly referred to as schema on read (Strang & Sun, 2017).

One of the outcomes of big data building is commonly known as "processing relocation from data to processing," which involves shifting the computation of data toward the location where the data is stored, rather than transferring the data to a separate processing location. The suggestion is that the volume of data is excessively large to be retrieved and transferred to an alternative resource for examination. Consequently, the analysis software is disseminated to the data-storing resources, and solely the outcomes are consolidated on a distinct resource. Given that data transfer rates are often constrained by I/O bandwidth, an alternative strategy would be to integrate query and filtering programs directly into the tangible storage medium (Kressbach, 2021).

The concept of Big Data primarily involves the expansion of data storage and analysis capabilities throughout horizontally-scaled resources. This bears resemblance to the approach taken by the compute-intensive modeling community in the past, where enormously parallel processing was adopted. Historically, traditional parallel computing programs employed a wide range of syncing and communication mechanisms and established varied communication structures. Presently, compared to the past, the size of data sets has grown significantly, reaching Petabyte as well as Exabyte levels. As a result, distributed computation frameworks have emerged, providing patterns such as map-reduce, which offer a dependable high-level, commercially feasible computing model based on commodities computing hardware, dynamic schedulers of resources, as well as synchronization methods. Big Data refers to a collection of data sets that possess certain characteristics, including size, speed, variety, variability, as well as veracity, which make it difficult to analyze via current or developed technologies and methods. This challenge is particularly relevant within a particular area of concern and at a specific point in time; additionally, the ultimate goal is to determine value from the data (Jeansoulin, 2016).

The aforementioned definition serves to differentiate Big Data from conventional transactional analyzing as well as BI, besides implying a diverse range of potential applications that encompass these domains. The primary objective of Big Data analysis is to extract distinctive value that is reliable, owing to the dependability of the underlying data. This is achieved by utilizing sophisticated analytical techniques on the entire dataset, irrespective of its magnitude. The act of parsing this objective aids in structuring the discourse on the significance of implementing Big Data applications.

1. The utilization of the complete corpus of pertinent data, instead of limited samples or subsets, across all scales of operations. The objective is to integrate decision-support time horizons, encompassing the past, present, as well as future, by utilizing statistically obtained insights from extensive data sets across all these dimensions.
2. The acquisition of reliable data is crucial for obtaining accurate insights. This can be achieved through either consolidating and

purifying deep data to establish a single form of reality or via utilizing statistical techniques to extract reliable information from large amounts of unrefined data (Hitzler & Janowicz, 2013).

3. The utilization of sophisticated analytical techniques and tools to extract insights and knowledge from data is referred to as sophisticated analytics. The utilization of diverse analytical and mining methodologies to extract insights via data patterns, including "long tail" evaluations, micro-segmentations, as well as others, can lead to expedited insights. However, such techniques may not be viable when limited by smaller amounts, slower speeds, narrower variations, as well as undetermined veracities.

An intricate inquiry pertains to the defining characteristics of "Big Data" and the threshold at which a dataset can be classified as such. The response to the question is inconclusive and contingent upon various factors. One aspect of this matter pertains to the relativity of the term "Big," given the increasing density of computation as well as storage capabilities, which includes the availability of greater capacity in more compact and effective form factors. As a result, what is currently deemed as "Big" may not retain the same designation in the future. The term "big data" is applied to datasets that exhibit improved company efficiency when processed using new scalable buildings, as compared to traditional architectures. Stated differently, the conventional relational database platform is incapable of attaining the said functionality (Kitchin, 2014).

The concept of big data primarily centers on the perspective that data is considered large due to the necessity for scalable frameworks to manage it. As a result, improved architectures with enhanced scalability have emerged to address the demands of big data (Pajić et al., 2018).

6.3. CLOUD COMPUTING

Cloud computing is a technology that provides a range of computing services, including storage devices, databases, networks, software, analytics, as well as intelligence, through the internet. This technology is designed to facilitate faster innovation, offer flexible resources, enable heavy computation, support parallel computation, as well as offer economies of scale. The abstraction of computing, storage, as well as network facilities to tasks as required, along with access to a plethora of prebuilt services, enables organizations to focus on their core business. The document presents a comprehensive overview of the distinct services provided by individual computing layers, along with a comparative analysis of their dissimilarities (Dillon et al., 2010).

The assortment of services for cloud computing that are currently accessible is extensive. However, the majority of them can be categorized into either of the groups that follow (Armbrust et al., 2010):

6.3.1. SaaS: Software as a Service

The (SaaS) sector is the most extensive cloud market and frequently employed business alternative in cloud computing services. Software as a Service (SaaS) is a software delivery model that enables the provision of applications to end-users via the Internet. Software as a Service (SaaS) apps are managed by external providers, and clients access their interfaces via web browsers. Because the majority of Software as a Service (SaaS) applications operate directly through a web browser, there is no requirement for the end-user to download or set up any software. SaaS vendors assume responsibility for managing applications, execution, data, middleware, operating systems, virtualization, storage facilities, servers, as well as networking. This comprehensive approach facilitates streamlined maintenance as well as assistance for enterprises (Godse & Mulik, 2009).

6.3.2. PaaS: Platform as a Service

The (PaaS) model facilitates the delivery of software as well as hardware via the internet, enabling developers to construct tailored applications. Platform as a Service (PaaS) facilitates the expeditious, uncomplicated, and economic development, testing, as well as installation of applications.

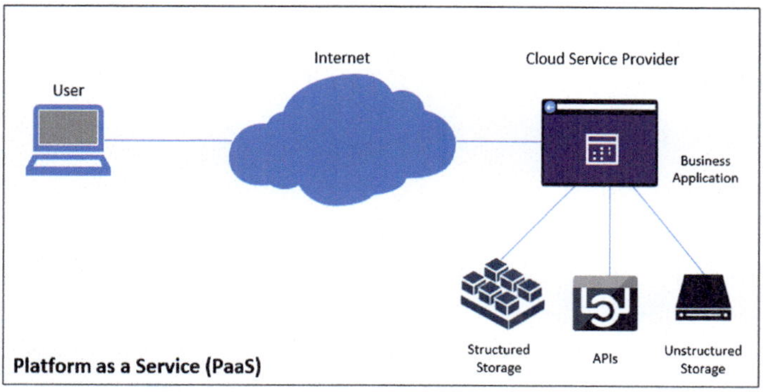

Figure 6.7. Illustration of platform as a service (PaaS).

Source: Francis Xavier Ovon, Creative Commons License.

This particular model enables businesses to develop and construct applications that are seamlessly integrated into Platform-as-a-Service (PaaS) software parts, while the management of the operating system, cloud computing, servers, data storage, network, as well as the PaaS software actually is handled by either enterprise operations or third-party providers. The aforementioned applications exhibit cloud characteristics, rendering them both highly available as well as scalable (Giessmann et al., 2012) (Figure 6.7).

6.3.3. IaaS: Infrastructure as a Service

(IaaS) refers to a cloud computing model where a third-party provider hosts virtualized computing resources over the internet. These resources can include servers, storage, networking, and other computing infrastructure components. Customers can then access and use these resources on a pay-per-use basis, allowing them to scale their infrastructure up or down as needed without having to invest in and maintain physical hardware. The cloud computing paradigm offers a self-service platform that enables organizations to obtain, monitor, and supervise remote data center infrastructures, including servers, storage devices, and networking services, via the use of virtualization technology. The responsibility of managing data, apps, runtime, middleware, as well as OS falls upon the users of Infrastructure as a Service (IaaS), while the providers are still accountable for managing virtualization, computers, hard drive retrieval, and networking. Infrastructure as a Service (IaaS) offers comparable functionalities to traditional data centers, while eliminating the need for physical maintenance (Manvi & Shyam, 2014).

6.4. CLOUD DATABASE PROVIDERS: AWS, AZURE, GOOGLE CLOUD

Cloud database suppliers are entities that provide cloud-oriented database solutions to both corporate and individual clients. The company offers managed database services that enable users to securely store, efficiently manage, and easily retrieve their data in a cloud-based environment. The three primary providers of cloud databases are Amazon's Web Services, Azure from Microsoft, as well as Google Cloud. This exposition will comprehensively examine each of these service providers, expounding on their provisions, characteristics, and benefits (Kaushik et al., 2021).

6.4.1. Amazon Web Services (AWS)

Amazon.com provides a comprehensive cloud-based computing platform known as AWS. AWS offers a range of database services that are tailored to meet diverse requirements. Several noteworthy database services are available by AWS:

a. Amazon RDS serves as a controlled database service offering support for multiple relational database platforms, including but not limited to PostgreSQL, MySQL, SQL Server, Oracle, as well as MariaDB. The software automates various database administration tasks that are typically time-consuming, including hardware provisioning, updating, backups, as well as failover. This enables users to direct their attention towards their applications (Hashemipour & Ali, 2020).

b. Amazon DynamoDB serves as a NoSQL database service that is fully managed and provided by AWS. The system exhibits high scalability and is specifically engineered to effectively manage substantial amounts of data while maintaining minimal latency. DynamoDB is a fitting choice for applications that necessitate adaptable, efficient, and expandable database functionalities.

c. Amazon Aurora represents a cloud-based relational database key designed specifically for cloud computing. This solution provides comparable efficiency and accessibility to that commercial-grade files while being significantly more cost-effective. Aurora has the capability to function seamlessly with both MySQL as well as PostgreSQL, thereby augmenting its proficiency and expandability (Mathew & Varia, 2014).

d. Amazon DocumentDB acts as a NoSQL record database service that has been fully managed and has compatibility with MongoDB. The technology in question presents the capacity for scalability, resilience, and availability, rendering it a fitting option for applications that prioritize document-oriented functionality.

e. AWS offers a diverse array of database services, including but not limited to Amazon Redshift, which is used for data storage, Amazon Neptune over databases with graphs, as well as Amazon ElastiCache over in-memory caching (Cook, 2018) (Figure 6.8).

Figure 6.8. Image of AWS.

Source: Amazon, Creative Commons License.

6.4.2. Microsoft Azure

The Microsoft Azure system is a solution for cloud computing offered by Microsoft. Azure provides a variety of database services that are tailored to accommodate diverse data types as well as workloads. Several notable database services offered by Azure are:

1. Azure SQL Database seems a relational database assistance that is managed and constructed on the foundation of SQL Server by Microsoft. The product provides elevated levels of efficiency, expandability, and protection. The Azure Microsoft SQL Database platform offers a range of capabilities such as automated tuning, integrated intelligence, and adaptable deployment alternatives (Lang et al., 2015).
2. Cosmos DB from Azure has become a database assistance that is distributed globally and supports multiple models. The system provides support for a variety of data models, such as key-value, record, column family, as well as time-series. Cosmos DB is a database management system that offers features such as automatic scaling, reliability, as well as low latency. These capabilities make it a fitting choice for applications that are distributed globally.
3. Azure Database over MySQL, as well as Azure Database over PostgreSQL, as well as Azure Database over PostgreSQL, are cloud-based services that provide completely managed databases to MySQL as well as PostgreSQL, respectively. The aforementioned features encompass high availability, computerized backups, as well as intelligent optimization of performance.

4. In addition to its primary offerings, Azure provides supplementary database services including Azure Synapse Analytics, which specializes in data storage, Azure Cache over Redis, which facilitates caching, along with Azure Database over MariaDB (Das et al., 2019).

6.4.3. Google Cloud

Google Cloud has become a cloud-based computing system provided by the technology giant, Google. Google Cloud offers a variety of database services that cater to a wide range of needs. Prominent database services offered by Google Cloud consist of:

a. Cloud SQL Services is a type of managed relationship-based database service with the capability to support both MySQL as well as PostgreSQL. The product provides automated backup functionality, robust availability features, and effortless scalability options. Cloud SQL is a suitable option for applications that necessitate conventional relational database functionalities.

b. Cloud Spanner seems a relational database assistance that is firmly consistent, horizontally adaptable, as well as globally distributed. The system offers transactions using ACID spanning multiple regions as well as features automated scaling as well as data synchronization capabilities. Cloud Spanner serves as a highly appropriate solution for apps that are critical to the success of an organization and necessitate global reach and robust consistency (Bisong & Bisong, 2019).

c. Cloud Firestore serves as a document-oriented NoSQL database assistance. The software offers instantaneous synchronization, offline functionality, and automated scalability. Firestore is a viable option for mobile and web-based applications that require instantaneous updates and the ability to function without an internet connection.

In addition to its primary offerings, Google Cloud provides supplementary database solutions such as BigQuery, which facilitates data warehousing, Cloud Bigtable, which enables efficient NoSQL, as well as Memorystore over Redis, which supports caching (Shah & Dubaria, 2019) (Figure 6.9).

Figure 6.9. Picture of logo of google cloud.

Source: Netscout, Creative Commons License.

6.4.4. Advantages of Cloud Database Providers

1. The capacity of cloud database suppliers enables users to adjust the size of their databases to meet their specific requirements. The attribute of scalability obviates the necessity for initial capital expenditure on hardware as well as empowers enterprises to proficiently manage fluctuating workloads (Avram, 2014).
2. Cloud databases have been designed to exhibit high availability by incorporating inherent redundancy and automated failover mechanisms. This guarantees the persistence of data accessibility in the face of potential hardware or network malfunctions.
3. Cloud database suppliers provide management services, which encompass a range of administrative duties including software updates, restores, and overall performance optimization. This enables users to allocate their time as well as finances more efficiently toward their primary business operations (Ahmadi & Aslani, 2018).
4. Cloud database vendors offer a global reach by means of data centers situated across various regions, which allows customers to store as well as imitate their data for enhanced performance along with data redundancy.
5. Cloud database practitioners employ stringent security protocols to safeguard data against unauthorized access as well as ensure adherence to industry norms and regulatory requirements. The provided security measures include encryption both at rest as well as during transit, authentication for users, along with access controls according to assigned roles.

In general, AWS, Azure, as well as Google Cloud offer extensive and dependable cloud-based database solutions, enabling enterprises to harness the benefits of cloud computing for their file storage and administration requirements (Abadi, 2009).

6.5. RELATIONSHIP BETWEEN THE CLOUD AND BIG DATA

The domains of Big Data along with Cloud Computing exhibit a strong interdependence. The concept of big data centers on the extraction of value, whereas cloud computing is centered around self-service models that are scalable, flexible, on-demand, along with pay-per-use. The analysis of big data necessitates significant on-demand computational capabilities along with dispersed storage. Cloud computing offers a seamless solution by providing elastic, incorporated computer resources, as well as the necessary storage and computational power to facilitate the analysis of big data. Cloud computing provides distributed processing capabilities that enable scalability as well as expansion via the use of virtual machines, thereby addressing the needs of rapidly increasing data volumes. The proliferation of analytical websites has arisen to meet the needs of users, particularly those of data-intensive enterprises, by furnishing contextualized analyzes of their stored data. The emergence of big data has led to the provision of cost-effective big data systems by service providers including Amazon, Microsoft, as well as Google. These systems are designed to capture data as well as incorporate analytics for offering proactive along with contextual experiences (Zanoon et al., 2017).

The utilization of cloud computing infrastructure represents a novel and highly effective method for delivering information technology services. The cloud computing environment comprises multiple providers as well as consumer terminals. The aforementioned encompasses a variety of computing devices and software, as well as services that are either pay-per-use or dependent on subscriptions, which are available online as well as within real-time. Big data resources are utilized for data collection, which is subsequently kept and analyzed in the cloud. Cloud computing offers readily available facilities and amenities that facilitate the uninterrupted management of data. The prevalent big data analytics designs include software-as-a-service (SaaS), platform-as-a-service (PaaS), and infrastructure-as-a-service (IaaS). Cloud-based analytics as well as Analytics just like a Service have emerged as contemporary solutions that cater to the needs of customers on a demand basis. Analytics as a Service (AaaS) is a cloud-based subscription model that offers data analysis tools and procedures (Figure 6.10).

Figure 6.10. Picture of analytical as a service (AaaS).

Source: Synoptec, Creative Commons License.

This approach provides a rapid and scalable method for integrating data in partially structured, unstructured, as well as structured formats, as well as transforming and analyzing them (Bahrami & Singhal, 2015).

Virtualization refers to the technique of generating a software-based or virtual depiction of network connections, operating systems, servers, storage, along with virtual desktops. The process involves the establishment of a simulated computing environment within a pre-existing server, which enables the execution of preferred software applications while ensuring that the server's additional features and services remain unaffected and available to other consumers. The practice of sharing a solitary physical instance for an item or application among numerous consumers and companies is commonly referred to as a method. This is accomplished through the allocation of a symbolic identifier to a tangible storage entity and the provision of a reference to that tangible resource upon request. The process of virtualization allows for the consolidation of numerous workloads, which are executed on various physical servers, and then transferred onto distinct virtual computers that are accommodated on a solitary physical server. Reducing the quantity of physical servers can lead to a decrease in data center intricacy, as well as a reduction in both capital as well as administrative expenses. This approach results in decreased resource utilization along with enhanced power efficiency in comparison to a setup involving multiple nodes. Virtual machines possess a high degree of flexibility and portability, thereby facilitating the transfer of workloads to alternative server hardware in the event of unforeseen downtime, equipment malfunction, or maintenance of applications. The utilization of computerized big data applications such as Hadoop offers advantages that are not attainable through physical structures in areas such as usage of resources, expenses, as well as data management. Online data encompasses a diverse array of data sources as well as enhances the accessibility of data across heterogeneous environments. Moreover, it facilitates the transmission of data at a rapid pace across the network, thereby expediting data processing (Hashem et al., 2015).

Preserving confidentiality and safeguarding the integrity of information constitute a crucial facet of big data. The hosting and processing of data on outside infrastructure as well as services within cloud computing presents a significant challenge with regard to security and confidentiality. The greater the magnitude, diversity, and accuracy of large-scale data, the more significant the potential hazards related to its confidentiality and protection. The utilization of mobile health technology has significantly transformed the delivery of healthcare services across various domains (Sandhu, 2021). The availability of diverse healthcare mobile apps has facilitated individuals in preserving their standard of living, physical well-being, medicinal guidance, and diagnosis. The transmission of data across mobile internet may present a significant risk to the privacy of information in the absence of adequate network security measures or in the event of unauthorized access. In order to instill confidence in users and ensure the protection of their data, it is imperative to uphold a range of new laws, confidentiality guidelines, regulations, safeguards, regulatory requirements, as well as service level commitments between suppliers and buyers. The safeguarding of big data within the cloud has become a crucial concern due to the necessity of shielding the data from malevolent external entities. Additionally, the secure management of vast storage capacity by cloud service providers (CSP) becomes of utmost importance (Marx, 2013).

The correlation within big data, as well as computing via the cloud adheres to the input-processing-output paradigm. The data input comprises voluminous data procured from diverse sources including smart devices like cellular phones, which may be structured, or partially structured in nature. The copious amount of data is subjected to a cleansing process and subsequently maintained in Hadoop or just other data repositories. The data that has been stored is subsequently subjected to cloud computing methods and devices in order to facilitate the provision of services. The computation steps encompass all the necessary tasks involved in converting the entered data. The output refers to the numerical or graphical representation that is obtained subsequent to the processing of data for the purposes of analyzing and displaying it (Zheng et al., 2013).

The (IoT) represents a shared characteristic across Cloud computing as well as big data. The voluminous data produced by Internet of Things (IoT) devices necessitates real-time analysis. CSP facilitate the transmission of data through the internet or leased lines and subsequently store it within data repositories. Cloud computing methodologies and instruments are subsequently employed to process and scrutinize the voluminous data housed in the cloud. It offers a means for information to traverse, be stored, and undergo analysis. Cloud computing offers a shared infrastructure for the integration of the IoT as well as big data. The domains of cloud computing, the IoT, as well as Big Data, exhibit a mutually reinforcing relationship (Khan et al., 2015) (Figure 6.11).

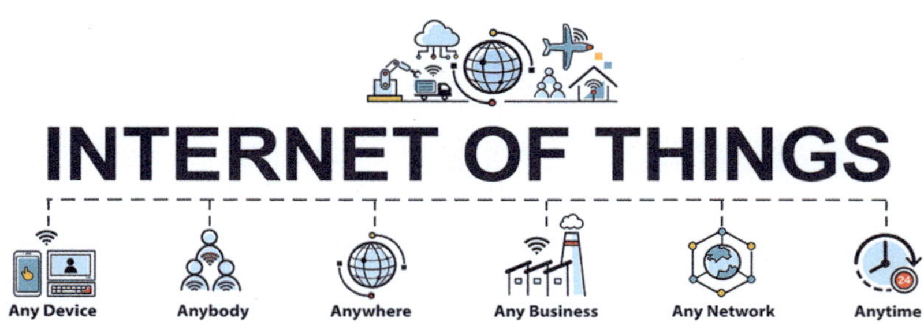

Figure 6.11. *Image of internet of things.*

Source: Muhammad AlYasfo, Creative Commons License.

6.6. SOLUTIONS AND METHODOLOGIES FOR DATA STORAGE

Various solutions have been proposed to address the storage and retrieval of vast quantities of data required by Big Data, with some of these solutions currently being implemented in Cloud computing. Internet-scale storage systems, like the Google Document System, endeavor to furnish the durability, expandability, and dependability that specific Internet services necessitate. Several alternative solutions offer object-store functionalities that enable the replication of files across various geographical locations to enhance redundancy, scaling, and data accessibility. Several cloud storage offerings are available, such as Amazon Simple Storage Service (S3), Nirvanix Dropbox, OpenStack Swift, as well as Microsoft Azure Binary Big Object (Blob) data storage. While these solutions may offer the necessary scalability as well as redundancy for numerous Cloud apps, they may not always satisfy the requirements for concurrency as well as efficiency in specific analytics applications (Jaseena & David, 2014).

Data location is a crucial factor in ensuring optimal performance over Big Data analytics apps. The reason for this is that the extensive amount of data implicated in the analytics renders it impractical to transmit the data for processing. In conventional systems, the favored approach involves executing CPU-intensive computations on an average to medium-scale data set. This is due to the fact that the data transmit-to-processing time ratio is relatively low, making it practical for transferring data to computer units. However, within the realm of Big Data, this methodology of relocating data to calculation nodes would result in a significant proportion of time being allocated to the transmission of data as opposed to processing time. Therefore, an alternative methodology is favored, whereby the computational process is relocated to the location of the data. The method of investigating data locality has been previously examined in scientific processes as well as Information Grids (Catthoor, 1999).

Within the realm of Big Data analysis, the MapReduce paradigm offers a compelling framework that leverages data locality to enhance application performance.

Hadoop, a freely available implementation based on the MapReduce framework, facilitates the formation of clusters that employ the HDFS, to divide and duplicate data sets across nodes, thereby enhancing the likelihood of their consumption by mappers. HDFS mitigates the effects of failures by duplicating datasets to a customizable quantity of nodes, while also capitalizing on the concurrence of numerous nodes. The development of a platform for analysis for processing Facebook's extensive data sets is being documented by. The platform employs the Scribe tool to consolidate logs generated by Web servers, subsequently exporting these into Hadoop Distributed File System (HDFS) archives. The analytics tasks are executed by a Hive-Hadoop group. The platform incorporates replication and compression methodologies, as well as columnar compression about Hive7, in order to efficiently store vast quantities of data (Lo et al., 2019).

Despite the prevalence of unorganized data in contemporary times, relational databases remain the preferred storage option for organizations seeking to manage information pertaining to customers, business operations, goods, and other relevant domains. As data stored in conventional DBMS grows older, it is transferred to information warehouses for the purpose of analysis and occasional retrieval. The suitability of models like MapReduce for analyzing relational data is generally limited. Hybrid solutions are being developed to integrate MapReduce functionality into the data analyzing as well as query operations of DBMS, as documented in reference.

The authors of present a design for databases that enables analytics through the use of SQL as well as MapReduce coding, while also facilitating the integration of diverse data sources within a DBMS. Several vendors for analytics and information mining options are adopting a strategy of deploying processing tasks in proximity to the information's storage location, leveraging models that include MapReduce. This approach aims to reduce redundancies in preparing the data, archiving, and processing. The processing and analytical capabilities associated with data are increasingly shifting in the direction of Enterprise Data Warehouses (EDWs) or being implemented in data hubs that allow efficient utilization across diverse data sets (Kumar et al., 2018).

A discernible pattern in the realm of Cloud computing involves the growing adoption of NoSQL databases which are the favored approach for information storage and retrieval. The NoSQL database management system employs a non-relational approach to storing data. According to Leavitt, unrelated models, including object-oriented, ordered, and graphed databases, have already been in existence for over half a century. However, in recent times, this approach has garnered increased attention with the emergence of models that include key-store, column-oriented, as well as document-based outlets. Levitt attributes the increase in interest to improved performance, capability to manage unstructured data, and compatibility with distributed environments.

The authors of conducted an in-depth examination of NoSQL databases, highlighting their strengths and weaknesses in the context of Cloud computing. The classification of NoSQL structures within the survey is based on how well they are able to handle distinct groups of consistency, accessibility, partitioning (CAP). The research study additionally investigates the supported data model of the NoSQL database systems under examination (Ranky, 2006).

6.7. DATA PROCESSING AND RESOURCE MANAGEMENT

In attempting to comprehend the notion of Big Data, one cannot evade the usage of terms that include "Map Reduce" as well as "Hadoop."

6.7.1. Hadoop

Hadoop comprises a Java-based coding framework that facilitates the computation of extensive datasets in the context of distributed computing and is available at no cost. This is a component of the Apache initiative that is financially supported via the Apache Software Foundation. The Hadoop cluster is organized in a Master/Slave architecture. Hadoop facilitates the processing of voluminous data sets across a network of servers, enabling the execution of apps on systems comprising numerous nodes as well as involving extensive storage capacity in the order of several terabytes. The utilization of a file system that is distributed within the Hadoop framework facilitates expeditious transmission of data as well as enables the system to maintain its standard functionality in the event of certain node failures. The aforementioned methodology mitigates the likelihood of a complete system breakdown, even in scenarios where several nodes can fail simultaneously. The Hadoop framework facilitates a technological solution that possesses the desirable attributes of scalability, cost-effectiveness, flexibility, and fault tolerance. The Hadoop Guidelines are utilized by prominent corporations such as Google, Amazon, as well as IBM, among others, to facilitate their applications that entail substantial quantities of data. The Hadoop framework comprises two primary sub-projects, namely MapReduce as well as (HDFS) (Luckow et al., 2016).

6.7.2. Map Reduce

The Hadoop MapReduce platform is utilized for developing applications that can efficiently handle massive data sets in a parallel manner across clusters of low-cost hardware resources while ensuring reliability and fault tolerance. The Map Reduce procedure initially partitions the data into discrete segments, which are subsequently subjected to parallel processing by Map tasks. Subsequently, the reducing tasks receive the results in the form of maps that have been sorted according to the structure. Typically, the data is entered and the results of the task are stored within a file system. It is responsible for managing the scheduling, tracking, and re-execution of failed tasks (Wang et al., 2013).

6.7.3. Hadoop Distributed File System (HDFS)

The Hadoop Distributed File System seems a distributed file system that provides data storage across all nodes within a Hadoop cluster. The process involves the integration of file systems present on individual nodes into a unified file system of significant magnitude. The High-Performance Distributed File System (HDFS) enhances reliability by implementing data replication across various sources to mitigate the impact of node failures (Borthakur, 2007) (Figure 6.12).

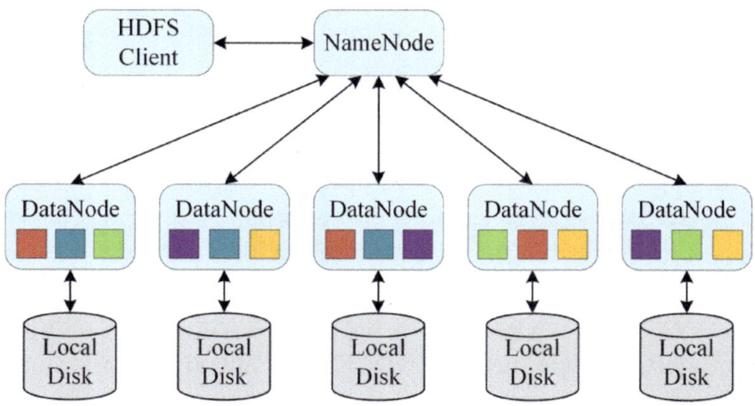

Figure 6.12. *Schematic of high-performance distributed file system (HDFS).*

Source: Forun Huawei, Creative Commons License.

6.8. ADVANTAGES OF BIG DATA APPLICATIONS

The term "big data application" pertains to distributed apps that operate on vast data sets. The task of exploring and analyzing data has become a challenging issue across various industries in the era of big data. The processing of vast and intricate datasets poses a challenge for conventional data processing apps, necessitating the creation of big data apps. The Google MapReduce platform as well as Apache Hadoop have become popular software systems for managing large-scale data applications, that often produce substantial volumes of intermediary data. The two primary domains of big data utilization are Manufacturing activities as well as Bioinformatics (Oussous et al., 2018).

The utilization of big data within the production sector offers a framework for promoting transparency. This technology has the potential to address uncertainties that arise from issues including erratic component functionality and accessibility. The theoretical framework for automated production in big data apps commences via the collection of data, which may encompass different kinds of sensory information, including but not limited to pressure, tremors, acoustics, electrical current, voltage, as well as controller data. The amalgamation of sensory information and historical records constitutes large-scale data within the realm of production. This amalgamation produces a substantial amount of data, commonly referred to as big data, which serves as the primary input for predictive techniques as well as preventative measures, including prognostics as well as medical management.

Hadoop has a significant application in the field of Bioinformatics, particularly in the areas of genome sequencing as well as other biological areas. Hadoop is utilized in the field of bioinformatics, which involves extensive analysis of vast amounts of data. The integration of computer clusters along with web-based interfaces via an accelerated shared computing structure is a characteristic feature of cloud computing (Muhammad et al., 2020).

Big data applications offer a diverse range of options and instruments that enable users to comprehensively map the information landscape for an organization. This facilitates the analysis of internal risks and dangers. The preservation of data is regarded as a primary benefit of big data. This approach enables one to identify confidential data that may not be adequately safeguarded and ensures its storage aligns with regulatory mandates.

Big data exhibits certain shared traits that are commonly observed.

1. The concept of big data involves the integration of data that is structured as well as unstructured.
2. The topic under consideration encompasses the aspects of speed and scaling, mobility, and safety, as well as adaptability and stability.
3. The prompt extraction of information is a crucial aspect in the realm of big data analytics, as it enables the extraction of value from diverse data sources like mobile devices, RF identification, the internet, as well as an expanding range of computerized sensory technologies (Bamiah et al., 2018).

Cloud storage can provide organizations and businesses with advantages such as enhanced speed, increased capacity, and improved scalability. Additionally, the data can be visualized by end users, while fresh business possibilities can be discovered by businesses. Big data offers a significant advantage in the form of data analytics, enabling the customization of website material and layout within real-time to cater to the unique preferences of every client visiting the site. The integration of big data and predictive analytics poses a significant challenge for numerous industries. The amalgamation leads to the investigation of the four domains:

1. Compute the potential hazards associated with extensive collections of assets.
2. The objective is to identify, mitigate, and conduct a subsequent examination of instances of fraudulent financial activity.
3. Enhance the process of collecting overdue payments.
4. Implement marketing campaigns with a high-value proposition (Correa, 2015).

6.9. CASE STUDIES

Numerous case studies have been conducted on the utilization of big data within cloud computing.

6.9.1. Tweet Mining in Cloud

The utilization of cloud computing was employed by Noordhuis et al. for the purpose of collecting and analyzing tweets. The calculations were executed utilizing the Amazon cloud computing platform. The process involved the crawling of tweets, followed by the application of a website's ranking method. Google utilizes Page Ranking as a metric to determine the significance of a website. In the context of the Website Ranking Algorithm, it can be observed that when the writer of a specific website A includes a hyperlink to a different website B, it's indicative of the writer's idea of the significance of website B. The significance of a particular page is determined by the number of in-edges it possesses. Therefore, the greater the quantity of links directed toward a webpage, the higher its perceived significance. The aforementioned principle is also applicable to tweets, wherein the significance of a tweet user is determined by the number of additional tweets they follow, rather than a webpage. It is necessary to perform a crawl of the Twitter interpersonal graph in order to calculate PageRank. The crawled data comprised approximately fifty million nodes as well as a total of 1.8 billion edges, thereby accounting for almost two-thirds of Twitter's projected user population (Alzahrani, 2018).

6.9.2. Redbus

Redbus is an internet-based travel agency that facilitates the booking of bus tickets in India. Redbus opted to utilize Google's data center for the purpose of enhancing sales to clients as well as managing their ticket reservations system through data handling and analysis. The utilization of Google BigQuery facilitated RedBus to swiftly handle extensive quantities of reservations as well as inventory information. The applications distributed across multiple servers facilitate the continuous streaming of customer queries, seat supplies, and booking details to a centrally controlled data-gathering system. The voluminous data gathered is subsequently transmitted to BigQuery, which executes intricate queries and expeditiously furnishes responses to diverse analytical inquiries, including the frequency with which a patron conducted searches for a particular location and encountered a paucity of bus services, all technical complications that may have arisen during the booking process, as well as the requisite team that ought to be notified. The comprehensive infrastructure facilitated RedBus in promptly resolving technical issues, mitigating revenue loss, and enhancing customer satisfaction (Scalici et al., 2014).

6.9.3. Nokia Mobile Company

Nokia cell phones are widely utilized for telecommunications by a significant number of individuals. Nokia collects substantial quantities of info on a petabyte scale via mobile devices to enhance its comprehension of customer interactions and optimize its customer experience. To formulate effective business decision-making strategies and gain a comprehensive understanding of user interactions, the organization has set up a Hadoop data center as a framework capable of storing the vast amount of unstructured information collected every day from mobile devices, services, log records, and various other sources. Through this configuration, Nokia was capable of executing intricate processing as well as computation using their extensive data, thereby obtaining further analytical insights into their user interactions, including "Which feature did they access subsequent to this one?" as well as "Where did they appear to experience difficulty?" The technology has facilitated the development of diverse applications, including 3D online maps that integrate traffic models capable of comprehending speed categories, past and present speeds on roadways, historical traffic patterns, elevation, ongoing incidents, video streams for the world, as well as additional features (Rajaniemi, 2003).

6.10. BIG DATA ISSUES

Despite the fact that big data addresses numerous contemporary challenges associated with processing large amounts of data, it remains a dynamic field that is continuously evolving and presents certain challenges. This section outlines certain unresolved concerns pertaining to big data as well as cloud computing (Kaisler et al., 2013).

With the exponential increase in data volume, retaining all data is not economically feasible. Hence, it is imperative for corporations to establish policies that delineate the life cycle as well as the termination date of data, which is commonly referred to as data governance. Furthermore, it is imperative to establish clear definitions regarding the individuals who are authorized to access their customer's data and the specific purposes for which such access is granted. The migration of data to cloud-based systems has prompted extensive research into the issues of security and privacy, which have emerged as significant concerns.

Database management systems for big data are commonly tasked with handling vast amounts of data from multiple origins, resulting in a variety of data types. Consequently, the issue of heterogeneity is now being investigated. Additional topics under investigation include disaster recovery, streamlined cloud data uploading, along with Exaflop computing (Katal et al., 2013).

This section presents an overview of the aforementioned issues.

6.10.1. Security

The subject of online computing as well as big data safety is presently a crucial area of research, as noted by Popović and Hocenski (2015). The matter of uploading information onto the internet for storage poses a concern for corporations. Interrogations pertaining to the veritable proprietorship of the data, its physical location, the individuals authorized to access it, and the nature of their entitlements present a challenging task in terms of explication. Enterprises intending to engage in operations via a cloud service provider ought to be cognizant of and inquire about the subsequent queries (Moreno et al., 2016).

The client of the cloud supplier remunerates for services and transfers their information onto the internet for storage. The question arises as to which of the 2 stakeholders truly owns the data. Furthermore, is it permissible for the service provider to utilize the data of the customer? What is the level of authorization granted and for what specific purposes may it be utilized? Will the cloud service provider derive advantages from the aforementioned data? In reality, it is imperative that the IT personnel tasked with the responsibility of upholding the client's data possess the necessary authorization to access information clusters. Consequently, it is advisable for clients to provide limited data access to ensure data security and restrict unauthorized personnel from accessing data without a valid reason. The aforementioned inquiries appear to be straightforward to address; however, it is imperative that they are comprehensively elucidated prior to engaging a particular service. The majority of security concerns often stem from internal sources within organizations. Therefore, it is prudent for companies to thoroughly scrutinize all data access regulations prior to finalizing a contractual agreement with a cloud service provider (Patil & Seshadri, 2014).

The legal status of private information may vary across different countries, necessitating an agreement on where it is stored to avoid potential legal repercussions for the client. It is important to note that information that is deemed lawful in one jurisdiction may be deemed unlawful in another, thereby exposing the customer to the risk of legal action.

The resolution of these inquiries is contingent on Service Level Agreements (SLAs), which necessitate meticulous scrutiny to comprehensively ascertain the responsibilities of every stakeholder as well as the scope of policies encompassed by the SLAs with respect to the company's data. Negotiation is a process that requires effective management.

With regard to the restriction of data access, Tu et al. (2013) and Popa et al. (2011) have proposed a viable method for encrypting data and performing analytical searches on encrypted information. By encrypting both data as well as queries, the issue of accessing information is effectively resolved. Despite its benefits, encryption incurs a cost that can result in increased query time during processing (Toshniwal et al., 2015).

6.10.2. Privacy

The act of collecting data and utilizing analytical instruments for the purpose of extracting information gives rise to various privacy-related apprehensions. The task of safeguarding data security as well as preserving privacy is growing increasingly challenging in light of the widespread dissemination and replication of information across the world. Analytics frequently extract sensitive user data, including medical records, usage of energy, browsing habits, and supermarket records. The disclosure of this data is subject to examination, which gives rise to apprehensions regarding the practice of profiling, discrimination, marginalization, and relinquishment of authority. Historically, organizations have employed diverse techniques for deidentification, such as anonymization or just encoding of data, in order to create a separation between data and actual identities (Salleh & Janczewski, 2016). Recent studies have demonstrated that the re-identification of anonymized data remains a possibility, thereby enabling the linkage of such data to particular people. One potential solution to address this issue involved treating all data to be individually identifiable as well as liable to rules and regulations. The act of discouraging organizations from utilizing de-identification techniques may result in heightened security and confidentiality hazards associated with data access.

The legal frameworks governing data security and confidentiality are founded on the notion of individual agency in relation to information, as well as upon principles like minimization along with the limitation of information as well as its purpose. However, the feasibility of adopting a privacy strategy that involves limiting the collection of information remains uncertain. Currently, privacy protocols for processing activities appear to rely on user approval and the intentional provision of data by individuals. The matter of privacy is unquestionably a concern that necessitates further enhancement, given that vast amounts of personal data are stored by systems on a daily basis (Smith et al., 2012).

6.10.3. Heterogeneity

The concept of big data encompasses not only large quantities of data additionally varying speeds of data acquisition based on source speeds of output as well as network latency, as well as a diverse range of data types. The latter encompasses vast and diverse datasets originating from multiple independent sources. According to Majhi and Shial (2015), diversity is a significant factor in the classification of big data. This is based on the notion that the accumulation of various types of data can be advantageous to both scientific and commercial endeavors (Zuech et al., 2015).

Big data database management systems receive data from diverse sources in varying formats as well as velocities. The reason for this phenomenon is that various information collectors exhibit a preference for their individual conceptual frameworks or procedures for the purpose of data recording. Additionally, the distinct nature of various applications leads to varied data representations, as stated by Wu et al. (2014). Managing diverse data types and varying velocity rates poses a formidable challenge that Big Data platforms must effectively address. The complexity of this endeavor is compounded by the incessant emergence of novel file formats lacking standardization. However, the task of establishing a uniform and comprehensive approach for representing and examining intricate and dynamic relationships derived from such data remains a formidable obstacle (Adnan & Akbar, 2019).

6.10.4. Data Governance

It is a veritable fact that the notion of storage being inexpensive, with the possibility of its cost decreasing even further, pertains to hardware pricing. Notwithstanding, a big data database management system (DBMS) also encompasses additional costs, including infrastructure upkeep, energy consumption, and software licensing fees. The aggregate of these expenditures constitutes the comprehensive cost of possession, which is approximated to be seven-fold greater than the expenses incurred in procuring the hardware (Fothergill et al., 2019). Given that the Total Cost of Ownership (TCO) exhibits a linear relationship with the expansion of big data, it is imperative to exercise stringent control over its growth. It is important to note that the concept of "Value" in the context of big data refers to the practice of storing only information that is deemed valuable, as vast quantities of data are rendered ineffective if they do not possess any inherent value. Data Governance has emerged as a solution to this issue by establishing policies that delineate the duration for which data remains viable. The notion encompasses a set of methodologies and administrative regulations that delineate the appropriate manner in which data ought to be handled throughout its economically valuable life span. These practices can be classified into three distinct categories (Yang et al., 2019):

1. According to Morgan Kaufmann (2013), fundamental practices involve the identification of important decision-makers, both within the IT as well as non-IT sectors, and their corresponding positions and duties with regard to ownership of data, value evaluation, as well as cost management.
2. Operational practices refer to the manner in which policies pertaining to the management of data are implemented. Commonly, these policies encompass a range of activities including but not limited to transferring data, keeping records, access privileges, allocation of costs, as well as backups and restoration (Tallon, 2013).

3. Relational practices refer to the formal description of the connections between the Chief Information Officer (CIO), executives in the company, as well as data users. These connections are established through various means such as sharing information, value assessment, education, training, as well as strategic information technology planning.

Data regulation is a broad concept that pertains to businesses that possess extensive datasets. It encompasses the formulation of guidelines for preserving valuable data and regulating accessibility for its entire life cycle. This matter necessitates careful consideration. In the absence of enforcement of governance policies, non-compliance is highly probable. While governing data can provide significant benefits, it is important to note that there are constraints to its value. In fact, overly stringent data governance measures can have adverse consequences besides a certain threshold (Kim & Cho, 2018).

6.10.5. Disaster Recovery

The significance of data in the business realm is considerable, as its loss can lead to a corresponding loss in value. In the event of unforeseen circumstances like earthquakes, floods, as well as fires, it is imperative to minimize data loss. In order to meet this stipulation, prompt accessibility of data while minimizing downtime along with loss is imperative in the event of any occurrence. Despite the significance of this matter, the current study in this specific domain is rather limited, as indicated by Subashini and Kavitha (2011), Wood et al. (2010), and Chang (2015).

It is crucial for large corporations to establish a disaster recovery program as a component of their data governance strategy. This plan should not solely rely upon backups of information to restore deleted information but also incorporate a series of protocols that facilitate the prompt substitute of the affected servers (Akter & Wamba, 2019).

The study outlined in Chang's (2015) work offers an appropriate method via a technical standpoint. The proposed approach is multi-purpose, enabling data restoration across multiple sites using various methods, resulting in a recovery rate of nearly 100%. According to the research, conventional data recovery techniques typically employ a "single-basket approach," wherein the restored information is obtained from a solitary source.

Efficient response to hazardous events is crucial as data loss can lead to financial losses. The efficacy of deploying large-scale data management systems in cloud environments while ensuring consistent availability and fault tolerance is significantly influenced by the implementation of disaster recovery methods (Yu et al., 2018).

6.10.6. Other Problems

The contemporary status of internet-based computing, big data, as well as big data platforms, gives rise to additional apprehensions. This section pertains to the transfer of data onto cloud-based platforms, the emergence of Exaflop calculating as a significant contemporary concern, and the challenges of scalability as well as flexibility in the context of online computing as well as big data (Madhavan & Richey, 2016).

a. The method of transferring information onto the internet is often characterized by a slow pace, prompting corporations to opt for the physical shipment of hard drives to data centers as a means of expediting data upload. Nonetheless, this approach is deemed impractical and unsafe for uploading information onto the internet. Over time, attempts have been made to enhance and develop effective data-uploading computations with the aim of reducing upload durations and ensuring secure data transfer to cloud storage. Nevertheless, this process continues to pose a significant bottleneck (Fan et al., 2014).

b. Exaflop computing has been a topic of extensive discourse in contemporary times, as evidenced by the works of Geller (2011) and Schilling (2014). Contemporary supercomputers along with cloud computing systems possess the capability to manage petabyte-scale data sets. Nevertheless, the management of exabyte-scale data sets remains a subject of significant concern, as the processing and transfer of such vast quantities of data necessitate high performance and bandwidth over network infrastructures. The efficacy of cloud computing as a solution may be called into question due to its perceived inferior speed in comparison to supercomputers, which can be attributed to the limitations imposed by existing bandwidth and latency. High-performance computing (HPC) systems are considered a highly viable solution, albeit one that comes with a substantial annual cost. Moreover, there exist multiple challenges in the development of exaflop high-performance computing systems, particularly with respect to optimizing power efficiency. In this context, the prevalent approach is to rely on GPU-based options as opposed to CPU-based ones. Issues pertaining to the requisite level of parallelism amongst a multitude of CPUs numbering in the hundreds of thousands are additionally present. The analysis of datasets within the Exabyte range necessitates the enhancement of big data & analytics, thereby presenting an additional challenge that remains unresolved (Stephens et al., 2015).

c. The topic of scalability as well as elasticity within cloud computing, specifically in relation to big databases, warrants additional investigation as current systems exhibit a limited capacity for automatically handling data surges. Typically, the act of scaling is initiated through manual means rather than automated processes. Current automatic scalable structures tend to employ reactive or just proactive algorithms, with a focus on enhancing performance when considering scalability. A system with flexibility that is deemed appropriate would facilitate proactive as well as reactive scaling, both manual along with automatic, by considering multiple dimensions including security, workload reset (i.e., the necessity to adjust the workload), as well as redundant service (which allows for failure tolerance as well as availability). In addition, extant algorithms for data rebalancing rely on the construction of histograms and the achievement of load adjustment. The latter mechanism guarantees a uniform allocation of workload across all servers. Nevertheless, the construction of histograms based on the load of individual servers incurs significant time as well as resource costs. As such, ongoing investigations are being carried out in this area to enhance these algorithms (Erikson, 2018).

6.10.7. Research Challenges

As previously mentioned, the integration of the cloud along with big data technologies exhibits a high level of compatibility. Despite the existing partnership between the aforementioned technologies, they still present certain challenges.

Table 6.1 presents a comprehensive overview of the current challenges associated with big data along with cloud computing. The initial column delineates the extant concerns, while the subsequent column expounds upon the current remedies. The remaining columns explicate the benefits and drawbacks of each respective solution (Anagnostopoulos et al., 2016).

Table 6.1. Tabular Representation of Big Data Issues.

Issues	Existent solutions	**Advantages**	**Disadvantages**
Security	Based on SLAs and data Encryption	Data is encrypted	Querying encrypted data is time consuming
Privacy	-De-identification -User consent	Provides a reasonable privacy or transfers responsibility to the user	It was proved that most deidentification mechanisms can be reverse engineered
Heterogeneity	One of the big data systems' characteristics is the ability to deal with different data coming at different velocities	The major types of data are covered up	It is difficult to handle such variety of data and such different velocities
Data Governance	Data governance documents	-Specify the way data is handled; -Specify data access policies; -Role specification; -Specify data life cycle	-The data life cycle is not easy to define; -Enforcing data governance policies so much can lead to counterproductive effects
Disaster recovery	Recovery plans	Specify the data recovery locations and procedures	Normally there is only one destination from which to secure data
Data Uploading	-Send HDDs to the cloud provider -Upload data through the Internet	Physically sending the data to the cloud provider is quicker than uploading data but it is much more unsecure	Physically sending data to the cloud provider is dangerous as HDDs can suffer damage from the trip. - Uploading data through the network is time-consuming and, without encryption, can be insecure
High Data processing (Exabyte datasets)	-Cloud computing -HPCs	Cloud computing is not so cost expensive as HPCs but HPCs are believed to handle Exabyte datasets much better	HPCs are very much expensive ant its total cost over a year is hard to maintain. On the other hand, cloud is believed that cannot cope with the requirements for such huge datasets

Scalability	Scalability exists at the three levels in the cloud stack. At the Platform level there is: horizontal (Sharding) and vertical scalability	Scalability allows the system to grow on demand	Scalability is mainly manual and is very much static. Most big data systems must be elastic to cope with data changes
Elasticity	There are several elasticity techniques such as Live Migration, Replication and Resizing	Elasticity brings the system the capability of accommodating data peaks	Most load variations assessments are manually made, instead of automatized

Source: Science direct, Creative Commons License.

While big data systems offer significant capabilities for enterprises and scientific research to gain insights from data, there exist certain concerns that warrant further investigation. It's imperative to allocate supplementary resources towards the enhancement of security protocols and the establishment of uniform data formats (Li et al., 2016). Scalability is a critical aspect of Big Data, and in commercial applications, it is predominantly achieved through manual rather than automatic means.

Additional investigation is necessary to address this issue. In relation to this specific domain, our intention is to employ flexible mechanisms to devise a resolution for the incorporation of scalability across multiple facets of large-scale data systems operating on cloud infrastructures. The objective is to examine the mechanisms that flexible software can employ to initiate scalability at various levels within the cloud stack. Therefore, the automatic as well as reactive accommodation of data peaks is facilitated.

This paper presents a summary of the integration of big information and cloud computing, emphasizing their benefits and demonstrating their compatibility. However, it also acknowledges the obstacles encountered by these technologies. With regard to the current issues, several potential advancements are identified for the upcoming years.

1. Data encryption is a potential solution for addressing safety and confidentiality concerns. Nonetheless, it is imperative for the latest systems to guarantee prompt access to data while minimizing the adverse impact of encryption upon processing times.
2. The issue of Big Data range can be effectively tackled through the implementation of data standardization techniques. It is our belief that this represents the subsequent measure toward mitigating the effects of heterogeneity.
3. The management and implementation of data security along with

data recovery strategies pose significant challenges. However, with the emergence of Big Data as a ubiquitous technology, organizations are beginning to recognize the necessity for these plans (Tsou, 2015).

4. In order to facilitate the process of uploading information onto the internet, it might be beneficial to implement novel and robust quality of services-based mechanisms.
5. The challenge of achieving exaflop computing necessitates government support and is deemed to be in the government's best interests. Currently, the most effective approaches involve the utilization of HPC systems and Graphics Processing Units (GPUs).
6. Various Big Data suppliers, including Amazon as well as Microsoft, widely employ scalability as well as elasticity methods. The primary focus pertains to the development of fully automated reactive as well as proactive systems which possess the ability to handle load specifications in an autonomous manner (Jin et al., 2015).

6.11. BIG DATA BUSINESS CHALLENGES

6.11.1. Utilities: Power Consumption Prediction

Smart meters are employed by utility companies to gage the consumption of power and gas. These technological instruments produce substantial amounts of data. The implementation of robust big data structures is imperative for the monitoring and analysis of energy production as well as consumption, which can be effectively achieved through the utilization of smart meters (Alahakoon & Yu, 2013).

6.11.2. Social Network: Sentiment Analysis

Social media enterprises, such as Twitter, must ascertain the discourse of their users and the prevalent subjects in order to conduct a comprehensive analysis of the prevailing sentiment.

6.11.3. Telecommunication: Predictive Analytics

The development of churn estimation techniques in the field of telecommunication is contingent upon profile information attributes. Through the analysis of subscribers' calling trends, predictive modeling has the capability to forecast churn (Wassouf et al., 2020).

6.11.4. Customer Service: Call Monitor

The utilization of application logs is a common practice in calling center big data applications with the purpose of enhancing performance. Prior to analysis, it is necessary to amalgamate log files that are in disparate formats.

6.11.5. Banking: Fraud Detection

Banking institutions ought to possess the capability to avert fraudulent activities on an individual transaction or account of an individual. It is imperative for big data options to conduct real-time analysis of transactions and furnish prompt suggestions for action while preventing fraudulent activities.

6.11.6. Retailers: Product Recommendation

Retailers have the capability to track and analyze users' browsing behavior and past purchases, enabling them to offer personalized product recommendations based on this information. It is imperative for sellers to provide confidentiality disclosures to consumers prior to the implementation of such applications (Dekimpe, 2020).

Tip

When performing in-memory analytics, it is essential to use distributed data storage .to optimize performance and handle large datasets efficiently

6.12. GOOD PRACTICES

The following are a set of effective design principles that can be applied to big data applications.

6.12.1. Good Architectural Design

The architecture for big data ought to offer cloud-based services that enable parallel as well as distributed processing. It is imperative that every node within the cluster exhibits independence, equality, and parallelism with respect to the remaining nodes. Sharing resources within such nodes would significantly diminish scalability.

The implementation of NoSQL databases can facilitate expedited retrieval of information and heightened performance. Both Lambda Kappa

Designs have the capability to perform processing tasks in both real-time as well as batch-processing modes (Villars et al., 2011).

6.12.2. Different Analytical Methods

Upon gathering data from various sources, organizations are required to determine the analytical approach that is most suitable for their business objectives. Several analytical techniques that are frequently employed include predictive analytics, flow analytics, preparing data, identification of fraud, and sentiment analysis. After determining the analytical approaches to be employed, big data applications must leverage data mining, deep learning, dispersed programming, statistical evaluation, memory-based analytics, as well as visualization techniques available via cloud computing (Sivarajah et al., 2017).

6.12.3. Use Appropriate Technique

It is not possible to employ a singular technique for the purpose of data analysis. Given that a significant portion of the data that is transmitted lacks structure, it is imperative to employ a suitable technological framework to effectively analyze this intricate data. In cases where immediate processing is required, Apache Spark may be utilized, whereas Hadoop may be employed for batch-processing needs. When there is a need to store information across multiple data centers, it becomes imperative to take into account the use of geographic databases. It's imperative to employ analytical tools which are well-suited for established databases (Wisesa et al., 2020).

6.12.4. Use In-Memory Analytics

The utilization of memory-based database analytics enables the execution of analytical operations in situ, where the data is stored. The system executes the necessary analytical operations during runtime. The co-location of requests and information within the identical server's memory results in a reduction in turnaround time (Garber, 2012).

6.12.5. Distributed Data Storage for In-Memory Analytics

In order to optimize the use of memory-based analytics, this is necessary to divide the data as well as maintain it in multiple data stores. The implementation of infrastructure for cloud computing presents a viable option for shared storage solutions that should be considered (Hu et al.,

2018).

6.12.6. Coordination Between Tasks and Data

In order to attain scalability along with fault tolerance, it is imperative to establish coordination within information and its corresponding processing operations. Specialized frameworks for managing clusters, like Zookeeper, have the potential to be utilized (Kimura et al., 2021).

Activity 6.1

Imagine you are working for a large retail company that wants to improve its product recommendation system. How would you utilize ETL processing with SQL and big data models to enhance the accuracy and effectiveness of the recommendation algorithm?

6.13. SUMMARY

This chapter explores ETL (Extract, Transform, Load) processing with SQL in the context of Big Data. It discusses the significance of Big Data models and their flexibility in handling structured, semi-structured, and unstructured data. The chapter also delves into cloud computing, including different service models and popular cloud database providers like AWS, Azure, and Google Cloud. It highlights the relationship between the cloud and Big Data, as well as solutions and methodologies for data storage and processing. The chapter concludes with an examination of the advantages, challenges, and business applications of Big Data, emphasizing the importance of good architectural design, analytical methods, and in-memory analytics.

Review Questions

1. What are the different types of data models used in Big Data processing?
2. Name three major cloud database providers and their advantages.
3. How does ETL processing with SQL contribute to data transformation and integration?
4. Explain the concept of schema-on-read and its significance in Big Data analytics.
5. What are the three categories of cloud computing services?

6. Describe the Hadoop framework and its role in data processing.
7. How does MapReduce facilitate distributed data processing?

REFERENCES

1. Abadi, D. J., (2009). Data management in the cloud: Limitations and opportunities. *IEEE Data Eng. Bull.*, *32*(1), 3–12.
2. Adnan, K., & Akbar, R., (2019). Limitations of information extraction methods and techniques for heterogeneous unstructured big data. *International Journal of Engineering Business Management*, *11*, 760–771.
3. Ahmadi, M., & Aslani, N., (2018). Capabilities and advantages of cloud computing in the implementation of electronic health record. *Acta Informatica Medica*, *26*(1), 24.
4. Akter, S., & Wamba, S. F., (2019). Big data and disaster management: A systematic review and agenda for future research. *Annals of Operations Research*, *283*, 939–959.
5. Alahakoon, D., & Yu, X., (2013). Advanced analytics for harnessing the power of smart meter big data. In: *2013 IEEE International Workshop on Intelligent Energy Systems (IWIES)* (Vol. 1, pp. 40–45). IEEE.
6. Albrecht, A., & Naumann, F., (2008). Managing ETL processes. *NTII*, *8*(2008), 12–15.
7. Alzahrani, S. M., (2018). Development of IoT mining machine for Twitter sentiment analysis: Mining in the cloud and results on the mirror. In: *2018 15th Learning and Technology Conference (L&T)* (Vol. 1, pp. 86–95). IEEE.
8. Anagnostopoulos, I., Zeadally, S., & Exposito, E., (2016). Handling big data: Research challenges and future directions. *The Journal of Supercomputing*, *72*, 1494–1516.
9. Armbrust, M., Fox, A., Griffith, R., Joseph, A. D., Katz, R., Konwinski, A., & Zaharia, M., (2010). A view of cloud computing. *Communications of the ACM*, *53*(4), 50–58.
10. Avram, M. G., (2014). Advantages and challenges of adopting cloud computing from an enterprise perspective. *Procedia Technology*, *12*, 529–534.
11. Awiti, J., Vaisman, A. A., & Zimányi, E., (2020). Design and implementation of ETL processes using BPMN and relational algebra.

Data & Knowledge Engineering, 129, 101837.

12. Bahrami, M., & Singhal, M., (2015). The role of cloud computing architecture in big data. *Information Granularity, Big Data, and Computational Intelligence, 5*, 275–295.
13. Bakshi, K., (2012). Considerations for big data: Architecture and approach. In: *2012 IEEE Aerospace Conference* (Vol. 1, pp. 1–7). IEEE.
14. Bamiah, S. N., Brohi, S. N., & Rad, B. B., (2018). Big data technology in education: Advantages, implementations, and challenges. *Journal of Engineering Science and Technology, 13*(2018), 229–241.
15. Bao, Y., Song, J., Leng, F., Wang, D., & Yu, G., (2007). Study and implementation of a new SQL-based ETL approach. *Wuhan University Journal of Natural Sciences, 12*(5), 804–808.
16. Bisong, E., & Bisong, E., (2019). An overview of google cloud platform services. *Building Machine Learning and Deep Learning Models on Google Cloud Platform: A Comprehensive Guide for Beginners*, 4,7–10.
17. Borthakur, D., (2007). The Hadoop distributed file system: Architecture and design. *Hadoop Project Website, 11*(2007), 21.
18. Catthoor, F., (1999). Energy-delay efficient data storage and transfer architectures and methodologies: Current solutions and remaining problems. *Journal of VLSI Signal Processing Systems for Signal, Image, and Video Technology, 21*, 219–231.
19. Chacko, A. M., Basheer, A. M., & Kumar, S. M., (2015). Capturing provenance for big data analytics done using SQL interface. In: *2015 IEEE UP Section Conference on Electrical Computer and Electronics (UPCON)* (Vol. 1, pp. 1–6). IEEE.
20. Cook, B., (2018). Formal reasoning about the security of amazon web services. In: *Computer Aided Verification: 30th International Conference, CAV 2018, Held as Part of the Federated Logic Conference, FloC 2018, Oxford, UK, July 14–17, 2018, Proceedings, Part I 30* (Vol. 2, pp. 38–47). Springer International Publishing.
21. Correa, F. R., (2015). Is BIM big enough to take advantage of big data analytics? In: *ISARC. Proceedings of the International Symposium on Automation and Robotics in Construction* (Vol. 32, pp. 1, 2). IAARC Publications.
22. Das, S., Grbic, M., Ilic, I., Jovandic, I., Jovanovic, A., Narasayya,

V. R., & Chaudhuri, S., (2019). Automatically indexing millions of databases in Microsoft azure SQL database. In: *Proceedings of the 2019 International Conference on Management of Data* (Vol. 1, pp. 666–679).

23. Davoudian, A., Chen, L., & Liu, M., (2018). A survey on NoSQL stores. *ACM Computing Surveys (CSUR)*, *51*(2), 1–43.
24. Dekimpe, M. G., (2020). Retailing and retailing research in the age of big data analytics. *International Journal of Research in Marketing*, *37*(1), 3–14.
25. Dillon, T., Wu, C., & Chang, E., (2010). Cloud computing: Issues and challenges. In: *2010 24th IEEE International Conference on Advanced Information Networking and Applications* (Vol. 2, pp. 27–33). IEEE.
26. Erikson, S. L., (2018). Cell phones≠ self and other problems with big data detection and containment during epidemics. *Medical Anthropology Quarterly*, *32*(3), 315–339.
27. Fan, J., Han, F., & Liu, H., (2014). Challenges of big data analysis. *National Science Review*, *1*(2), 293–314.
28. Fothergill, B. T., Knight, W., Stahl, B. C., & Ulnicane, I., (2019). Responsible data governance of neuroscience big data. *Frontiers in Neuroinformatics*, *13*, 28.
29. Galici, R., Ordile, L., Marchesi, M., Pinna, A., & Tonelli, R., (2020). Applying the ETL process to blockchain data. prospect and findings. *Information*, *11*(4), 204.
30. Garber, L., (2012). Using in-memory analytics to quickly crunch big data. *Computer*, *45*(10), 16–18.
31. Giessmann, A., & Stanoevska-Slabeva, K., (2012). Business models of platform as a service (PaaS) providers: Current state and future directions. *JITTA: Journal of Information Technology Theory and Application*, *13*(4), 31.
32. Godse, M., & Mulik, S., (2009). An approach for selecting software-as-a-service (SaaS) product. In: *2009 IEEE International Conference on Cloud Computing* (Vol. 2, pp. 155–158). IEEE.
33. Hashem, I. A. T., Yaqoob, I., Anuar, N. B., Mokhtar, S., Gani, A., & Khan, S. U., (2015). The rise of "big data" on cloud computing: Review and open research issues. *Information Systems*, *47*, 98–115.
34. Hashemipour, S., & Ali, M., (2020). Amazon web services (AWS)–an overview of the on-demand cloud computing platform. In: *Emerging*

Technologies in Computing: Third EAI International Conference, iCETiC 2020, London, UK, August 19–20, 2020, Proceedings 3 (Vol. 3, pp. 40–47). Springer International Publishing.

35. Hitzler, P., & Janowicz, K., (2013). Linked data, big data, and the 4th paradigm. *Semantic Web*, *4*(3), 233–235.
36. Hu, F., Yang, C., Schnase, J. L., Duffy, D. Q., Xu, M., Bowen, M. K., & Song, W., (2018). ClimateSpark: An in-memory distributed computing framework for big climate data analytics. *Computers & Geosciences*, *115*, 154–166.
37. Janković, S., Mladenović, S., Mladenović, D., Vesković, S., & Glavić, D., (2018). Schema on read modeling approach as a basis of big data analytics integration in EIS. *Enterprise Information Systems*, *12*(8, 9), 1180–1201.
38. Jaseena, K. U., & David, J. M., (2014). Issues, challenges, and solutions: Big data mining. *CS & IT-CSCP*, *4*(13), 131–140.
39. Jeansoulin, R., (2016). Review of forty years of technological changes in geomatics toward the big data paradigm. *ISPRS International Journal of Geo-Information*, *5*(9), 155.
40. Jin, X., Wah, B. W., Cheng, X., & Wang, Y., (2015). Significance and challenges of big data research. *Big Data Research*, *2*(2), 59–64.
41. Kaisler, S., Armour, F., Espinosa, J. A., & Money, W., (2013). Big data: Issues and challenges moving forward. In: *2013 46th Hawaii International Conference on System Sciences* (Vol. 1, pp. 995–1004).
42. Katal, A., Wazid, M., & Goudar, R. H., (2013). Big data: Issues, challenges, tools, and good practices. In: *2013 Sixth International Conference on Contemporary Computing (IC3)* (Vol. 1, pp. 404–409). IEEE.
43. Kaushik, P., Rao, A. M., Singh, D. P., Vashisht, S., & Gupta, S., (2021). Cloud computing and comparison based on service and performance between amazon AWS, Microsoft azure, and google cloud. In: *2021 International Conference on Technological Advancements and Innovations (ICTAI)* (Vol. 1, pp. 268–273). IEEE.
44. Khan, Z., Anjum, A., Soomro, K., & Tahir, M. A., (2015). Towards cloud based big data analytics for smart future cities. *Journal of Cloud Computing*, *4*(1), 1–11.
45. Kim, H. Y., & Cho, J. S., (2018). Data governance framework for

big data implementation with NPS case analysis in Korea. *Journal of Business and Retail Management Research, 12*(3), 98–100.

46. Kimura, A., Yokozawa, T., & Ozaki, H., (2021). Clarifying the biomechanical concept of coordination through comparison with coordination in motor control. *Frontiers in Sports and Active Living, 3*, 753062.
47. Kitchin, R., (2014). Big data, new epistemologies, and paradigm shifts. *Big Data & Society, 1*(1), 8481–8489.
48. Kressbach, M., (2021). Period hacks: Menstruating in the big data paradigm. *Television & New Media, 22*(3), 241–261.
49. Kumar, P. R., Raj, P. H., & Jelciana, P., (2018). Exploring data security issues and solutions in cloud computing. *Procedia Computer Science, 125*, 691–697.
50. Lang, W., Bertsch, F., DeWitt, D. J., & Ellis, N., (2015). Microsoft azure SQL database telemetry. In: *Proceedings of the Sixth ACM Symposium on Cloud Computing* (Vol. 2, pp. 189–194).
51. Lee, S., & Park, S., (2021). Performance analysis of big data ETL process over CPU-GPU heterogeneous architectures. In: *2021 IEEE 37th International Conference on Data Engineering Workshops (ICDEW)* (Vol. 2, pp. 42–47). IEEE.
52. Li, S., Dragicevic, S., Castro, F. A., Sester, M., Winter, S., Coltekin, A., & Cheng, T., (2016). Geospatial big data handling theory and methods: A review and research challenges. *ISPRS Journal of Photogrammetry and Remote Sensing, 115*, 119–133.
53. Lo, S. K., Liu, Y., Chia, S. Y., Xu, X., Lu, Q., Zhu, L., & Ning, H., (2019). Analysis of blockchain solutions for IoT: A systematic literature review. *IEEE Access, 7*, 58822–58835.
54. Luckow, A., Paraskevakos, I., Chantzialexiou, G., & Jha, S., (2016). Hadoop on HPC: Integrating Hadoop and pilot-based dynamic resource management. In: *2016 IEEE International Parallel and Distributed Processing Symposium Workshops (IPDPSW)* (Vol. 2, pp. 1607–1616). IEEE.
55. Madhavan, K., & Richey, M. C., (2016). Problems in big data analytics in learning. *Journal of Engineering Education, 105*(1), 6–14.
56. Manvi, S. S., & Shyam, G. K., (2014). Resource management for Infrastructure as a Service (IaaS) in cloud computing: A survey. *Journal of Network and Computer Applications, 41*, 424–440.

57. Marx, V., (2013). The big challenges of big data. *Nature*, *498*(7453), 255–260.
58. Mathew, S., & Varia, J., (2014). *Overview of Amazon Web Services* (Vol. 105, pp. 1–22). Amazon Whitepapers.
59. Moreno, J., Serrano, M. A., & Fernández-Medina, E., (2016). Main issues in big data security. *Future Internet*, *8*(3), 44.
60. Muhammad, R. N., Tasmin, R., & Aziati, A. N., (2020). Sustainable competitive advantage of big data analytics in higher education sector: An overview. In: *Journal of Physics: Conference Series* (Vol. 1529, p. 042100). IOP Publishing.
61. Oussous, A., Benjelloun, F. Z., Lahcen, A. A., & Belfkih, S., (2018). Big data technologies: A survey. *Journal of King Saud University-Computer and Information Sciences*, *30*(4), 431–448.
62. Pajić, V., Govedarica, M., & Amović, M., (2018). Model of point cloud data management system in big data paradigm. *ISPRS International Journal of Geo-Information*, *7*(7), 265.
63. Patgiri, R., & Ahmed, A., (2016). Big data: The v's of the game changer paradigm. In: *2016 IEEE 18th International Conference on High Performance Computing and Communications; IEEE 14th International Conference on Smart City; IEEE 2nd International Conference on Data Science and Systems (HPCC/SmartCity/DSS)* (Vol. 1, pp. 17–24). IEEE.
64. Patil, H. K., & Seshadri, R., (2014). Big data security and privacy issues in healthcare. In: *2014 IEEE International Congress on Big Data* (Vol. 1, pp. 762–765). IEEE.
65. Rajaniemi, M., (2003). Nokia mobile phones' development of business logistics and customer support. In: *Information Technology Enabled Global Customer Service* (Vol. 1, pp. 249–257). IGI Global.
66. Ranky, P. G., (2006). An introduction to radio frequency identification (RFID) methods and solutions. *Assembly Automation*, *26*(1), 28–33.
67. Rodriguez-Muro, M., Hardi, J., & Calvanese, D., (2012). Quest: Efficient SPARQL-to-SQL for RDF and OWL. *Proceedings of the ISWC 2012 Posters & Demonstrations Track: ISWC 2012 Posters & Demos, Collected Abstracts, November 11–15, 2012* (Vol. 914, pp. 53–56).
68. Salleh, K. A., & Janczewski, L., (2016). Technological, organizational, and environmental security and privacy issues of big data: A literature

review. *Procedia Computer Science*, *100*, 19–28.

69. Sandhu, A. K., (2021). Big data with cloud computing: Discussions and challenges. *Big Data Mining and Analytics*, *5*(1), 32–40.
70. Scalici, S. P., Carry, Sr. W., Donnell, E., Rado, Z., & Hartmann, M. R., (2014). Red bus lane treatment evaluation. In: *T&DI Congress 2014: Planes, Trains, and Automobiles* (Vol. 1, pp. 132–141).
71. Shah, J., & Dubaria, D., (2019). Building modern clouds: Using docker, Kubernetes & Google cloud platform. In: *2019 IEEE 9th Annual Computing and Communication Workshop and Conference (CCWC)* (Vol. 2, pp. 0184–0189). IEEE.
72. Sharma, S., Tim, U. S., Wong, J., Gadia, S., & Sharma, S., (2014). A brief review on leading big data models. *Data Science Journal*, *13*, 138–157.
73. Silva, Y. N., Almeida, I., & Queiroz, M., (2016). SQL: From traditional databases to big data. In: *Proceedings of the 47th ACM Technical Symposium on Computing Science Education* (Vol. 1, pp. 413–418).
74. Sivarajah, U., Kamal, M. M., Irani, Z., & Weerakkody, V., (2017). Critical analysis of big data challenges and analytical methods. *Journal of Business Research*, *70*, 263–286.
75. Smith, M., Szongott, C., Henne, B., & Von, V. G., (2012). Big data privacy issues in public social media. In: *2012 6th IEEE International Conference on Digital Ecosystems and Technologies (DEST)* (Vol. 1, pp. 1–6). IEEE.
76. Stephens, Z. D., Lee, S. Y., Faghri, F., Campbell, R. H., Zhai, C., Efron, M. J., & Robinson, G. E., (2015). Big data: Astronomical or genomical? *PLoS Biology*, *13*(7), e1002195.
77. Strang, K. D., & Sun, Z., (2017). Big data paradigm: What is the status of privacy and security? *Annals of Data Science*, *4*, 1–17.
78. Sun, Z., Strang, K. D., & Pambel, F., (2018). Privacy and security in the big data paradigm. *Journal of Computer Information Systems, 4*(1), 45–49.
79. Tallon, P. P., (2013). Corporate governance of big data: Perspectives on value, risk, and cost. *Computer*, *46*(6), 32–38.
80. Toshniwal, R., Dastidar, K. G., & Nath, A., (2015). Big data security issues and challenges. *International Journal of Innovative Research in Advanced Engineering (IJIRAE)*, *2*(2), 12–16.

81. Tsai, C. W., Lai, C. F., Chao, H. C., & Vasilakos, A. V., (2015). Big data analytics: A survey. *Journal of Big Data*, *2*(1), 1–32.
82. Tsou, M. H., (2015). Research challenges and opportunities in mapping social media and big data. *Cartography and Geographic Information Science*, *42*(1), 70–74.
83. Uddin, M. F., & Gupta, N., (2014). Seven V's of big data understanding big data to extract value. In: *Proceedings of the 2014 Zone 1 Conference of the American Society for Engineering Education* (Vol. 1, pp. 1–5). IEEE.
84. Villars, R. L., Olofson, C. W., & Eastwood, M., (2011). *Big Data: What it is and Why You Should Care* (Vol. 14, pp. 1–14). White Paper, IDC.
85. Wang, L., Tao, J., Ranjan, R., Marten, H., Streit, A., Chen, J., & Chen, D., (2013). G-Hadoop: MapReduce across distributed data centers for data-intensive computing. *Future Generation Computer Systems*, *29*(3), 739–750.
86. Wassouf, W. N., Alkhatib, R., Salloum, K., & Balloul, S., (2020). Predictive analytics using big data for increased customer loyalty: Syriatel telecom company case study. *Journal of Big Data*, *7*, 1–24.
87. Wisesa, O., Adriansyah, A., & Khalaf, O. I., (2020). Prediction analysis for business to business (B2B) sales of telecommunication services using machine learning techniques. *Majlesi Journal of Electrical Engineering*, *14*(4), 145–153.
88. Yang, L., Li, J., Elisa, N., Prickett, T., & Chao, F., (2019). Towards big data governance in cybersecurity. *Data-Enabled Discovery and Applications*, *3*, 1–12.
89. Yu, M., Yang, C., & Li, Y., (2018). Big data in natural disaster management: A review. *Geosciences*, *8*(5), 165.
90. Zanoon, N., Al-Haj, A., & Khwaldeh, S. M., (2017). Cloud computing and big data is there a relation between the two: A study. *International Journal of Applied Engineering Research*, *12*(17), 6970–6982.
91. Zheng, Z., Zhu, J., & Lyu, M. R., (2013). Service-generated big data and big data-as-a-service: An overview. In: *2013 IEEE International Congress on Big Data* (Vol. 1, pp. 403–410). IEEE.
92. Zuech, R., Khoshgoftaar, T. M., & Wald, R., (2015). Intrusion detection and big heterogeneous data: A survey. *Journal of Big Data*, *2*(1), 1–41.

CHAPTER

7

Cloud Databases and Database-as-a-Service (DBaaS)

CONTENT

7.1. INTRODUCTION

Cloud databases are actually databases that are maintained and operated within the cloud computing infrastructure. Cloud-based systems and amenities are utilized to offer a platform for the storage, management, and accessibility of data. Cloud databases present a compelling option for contemporary data management requirements due to their capacity for scalability, flexibility, along with simplified administration, in contrast to conventional on-premises databases (Lehner et al., 2010).

The notion of cloud databases is centered on the utilization of the extensive computational capabilities of the cloud to furnish dependable, readily accessible, and economical database solutions. Cloud-based databases obviate the necessity for enterprises to allocate resources towards procuring and sustaining their own tangible hardware infrastructure, a process that can be both costly and labor-intensive. Alternatively, organizations can utilize the infrastructure of cloud service providers (CSP) to store and oversee their data.

Cloud-based databases can be classified into various categories by the kind of information model they endorse. These categories comprise relational, NoSQL, as well as NewSQL databases. Cloud-based relational databases, including Amazon RDS as well as Azure SQL Database, provide a means of storing structured data while also facilitating SQL queries as well as ACID transactions (Figure 7.1).

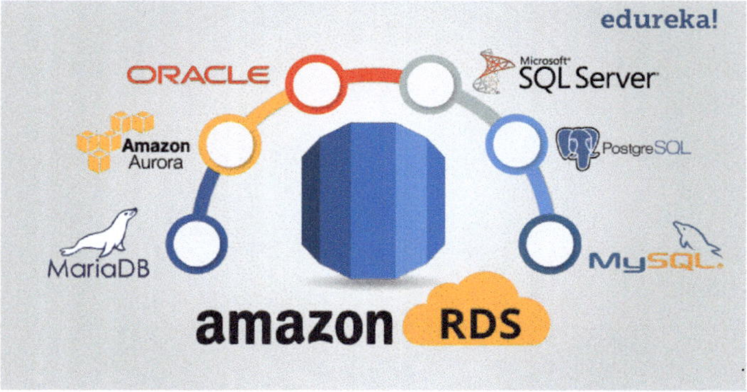

Figure 7.1. Image of Amazon RDS elements.

Source: Scandiplan, Creative Commons License.

NoSQL cloud databases including Amazon DynamoDB as well as Google Cloud Fire Store have been developed to provide adaptable and expandable storage solutions for unorganized or partially organized data. Cloud-based NewSQL databases, exemplified by CockroachDB as well as Google Cloud Spanner, amalgamate the scalability attributes of NoSQL database systems with the consistency as well as relational functionalities of conventional SQL databases (Munir, 2015) (Figure 7.2).

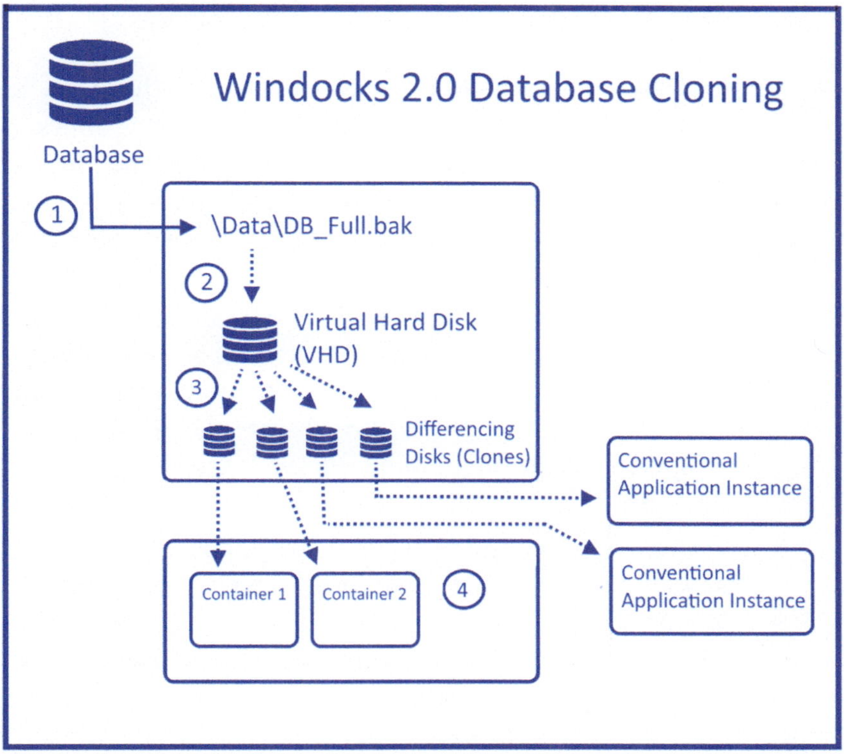

Figure 7.2. Illustration of SQL server container.

Source: Expert exchange, Creative Commons License.

Cloud-based databases and the provision of Database-as-a-Service (DBaaS) offer a multitude of advantages to entities of varying magnitudes. Several notable benefits can be identified:

1. Cloud databases provide the capability to adjust the scale of resources in response to varying levels of demand. Organizations possess the capability to flexibly allocate resources, including storage as well as computing power, in order to accommodate variable workloads without experiencing any downtime or incurring substantial infrastructure expenses.

2. Cloud databases have been cost-effective due to their pay-as-you-go model, which enables organizations to avoid significant initial investments within hardware as well as infrastructure. A pay-per-use model is a cost-efficient option, particularly for small enterprises or newly established firms with financial constraints, as they are charged solely for the resources they consume (Li et al., 2022).
3. Cloud databases offer the capacity to promptly allocate and launch novel databases or modify pre-existing ones, thus exhibiting flexibility and agility. The attribute of agility empowers organizations to swiftly respond to changing business requirements and expedite the deployment of novel applications or services.
4. Cloud databases are engineered to possess high availability, incorporating inherent redundancy as well as failover mechanisms for fault tolerance. Replication of data is a common practice across numerous geographically dispersed data centers, with the aim of guaranteeing uninterrupted business operations and mitigating the likelihood of data loss.
5. Cloud databases commonly provide automated backup and disaster recovery functionalities, thereby decreasing the administrative workload and guaranteeing the persistence of data. In the event of inadvertent deletions, corruptions, or just other catastrophic incidents, entities can readily reinstate data to a prior state (Al-Refai et al., 2021).
6. Cloud database service providers employ stringent security measures to safeguard data, such as encryption during data transmission and storage, access management, and frequent security patches to ensure compliance. Certification and audit procedures for compliance are frequently implemented to fulfill regulatory obligations that are specific to the industry.

7.2. EVOLUTION OF DBAAS IN THE CLOUD COMPUTING LANDSCAPE

The evolution of DBaaS has progressed in tandem with the swift expansion of cloud computing. At the outset, CSP placed their primary emphasis on furnishing infrastructure services including virtual machines as well as storage. The emergence of DBaaS can be attributed to the increasing demand for directed databases, which was seen as a logical progression of cloud offerings (Botta et al., 2016).

In the initial stages of the DBaaS provision, the scope of services was confined to fundamental database administration functions, such as the allocation of resources and data backup (Figure 7.3).

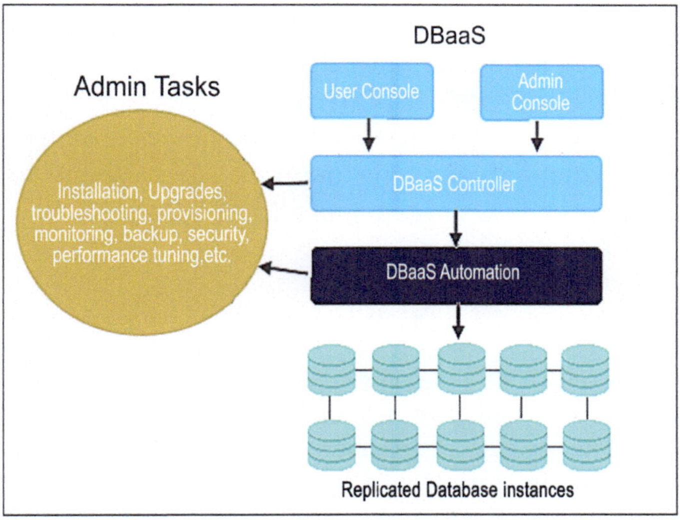

Figure 7.3. Schematic of private DBaaS.

Source: Paul Stanton, Creative Commons License.

In the course of time, service providers have progressively introduced more sophisticated functionalities and capabilities, such as automated scaling, replication, as well as monitoring. The process of evolution has facilitated the delegation of numerous intricate database administration responsibilities to the cloud provider by organizations, thereby allowing them to concentrate on their fundamental business operations.

Advancements regarding the number of supported database models have also been observed in the field of DBaaS. In the past, relational databases held a dominant position in the market. However, the emergence of big data as well as web-scale applications led to the increased popularity of NoSQL databases. CSP have taken measures to address the varied data storage requirements of contemporary applications by providing fully controlled NoSQL database services (Difallah et al., 2013). Moreover, the incorporation of cloud-native technology, including serverless computing as well as containerization, has proven advantageous for DBaaS. These technologies facilitate the development and implementation of applications by organizations, which can effectively interface with cloud-based databases, thereby capitalizing on the cloud's adaptable and scalable infrastructure. With the ongoing evolution of cloud computing, it is anticipated that DBaaS will undergo further advancements and become increasingly sophisticated. Current developments in the field involve the incorporation of machine learning and AI functionalities into cloud-based databases, enhanced data governance as well as privacy attributes, and more seamless connectivity with edge servers and Internet of Things (IoT) devices. In general, the development of DBaaS has revolutionized the way entities undertake database administration, providing unparalleled scalability, adaptability, and ease in the epoch of cloud computing Design Considerations for Cloud Databases (Sharma, 2016).

7.2.1. Scalability and Elasticity

The consideration of scalability as well as elasticity is of utmost importance in the design of a cloud-based database. They make sure the database is capable of efficiently managing increasing workloads and adapting to varying levels of demand. Within cloud computing environments, there exist two predominant methods of scaling: horizontal scaling as well as vertical scaling.

Horizontal scaling, which is also referred to as scaling out, pertains to the distribution of a database workload throughout numerous nodes or servers. The expansion of nodes within the system enables the database to effectively manage heightened levels of traffic and workload. The utilization of multiple servers for workload distribution is an approach that enhances performance, leading to improved resource utilization as well as increased throughput (Ferretti et al., 2013).

Cloud-based databases frequently offer functionalities and mechanisms that facilitate horizontal scaling in an effortless way. Controlled database services are provided by certain cloud providers, which facilitate the automatic distribution of data throughout multiple instances or just servers. This obviates the necessity for manually configuring and guarantees that the process of scaling remains unobtrusive to the application. Auto-scaling is a critical functionality in cloud computing environments, as it enables databases to automatically adjust to changes in workload. Auto-scaling enables the database to adapt its resources in response to varying levels of demand, thereby ensuring optimum efficiency as well as cost-effectiveness (Achar, 2019; Zhang et al., 2021;).

The implementation of auto-scaling methodologies entails the systematic observation of critical indicators, including processor utilization, utilization of memory, or just incoming asked rate. Upon reaching predetermined thresholds, the system initiates automatic provisioning of additional resources or scaling down of existing ones. The adaptability of the database guarantees its capacity to manage surges in workload without excessively allocating resources during intervals of reduced demand.

CSP present a range of auto-scaling functionalities, including AWS Auto Scaling or just Azure Autoscaling, that enable the user to establish scaling policies and regulations based on predetermined criteria. The formulation of policies can be predicated upon temporal schedules, pre-established metrics, or individually tailored metrics (Agrawal et al., 2011).

7.2.2. Availability and Fault Tolerance

The maintenance of high availability as well as fault tolerance has the utmost importance for cloud databases in order to mitigate the occurrence of system failures and information loss. Cloud-based databases utilize various techniques such as replication, dissemination of information, and detection of failures along with recovery mechanisms to accomplish their objectives.

The topics of replication as well as information distribution are being discussed. The process of replication entails generating numerous duplicates of a given database across distinct servers or geographical locations. The implementation of redundancy guarantees the availability of data from alternative replicas in the event of a server or regional malfunction. The process of replication can manifest in either synchronous or asynchronous modes (Verbitski et al., 2017).

Synchronous replication is a mechanism that guarantees the uniform application of modifications across all replicas prior to the finalization of a transaction. Although this approach enhances the reliability of data consistency, it may result in increased latency due to the need for extra network round trips. In contrast, asynchronous replication enables improved performance by permitting replicas to trail marginally behind the original database. The aforementioned methodology prioritizes enhanced response times over complete uniformity. Cloud-based databases frequently offer integrated replication mechanisms that enable the configuration of replication across numerous availability areas or regions. Geographically distributing data aids in mitigating the potential for data loss or lack of availability resulting from a singular point of failure (Radhakrishnan, 2012).

Cloud databases are equipped with mechanisms that enable the detection of failures and prompt recovery from them. The process of detecting failures entails the continuous monitoring of the well-being and condition of file nodes or situations. In the event of a detected failure, the system has the capability to commence the recovery process through automated node restarts or rerouting of traffic to functional replicas.

The process of recovering from failure can be accomplished by utilizing various techniques, including automatic failover, which involves the transfer of responsibilities from a failed node to another standby node, or via the implementation of load-balancing processes that effectively distribute traffic throughout all available nodes. The methodologies aim to mitigate the adverse effects of system malfunctions and guarantee uninterrupted accessibility.

Furthermore, cloud-based databases frequently offer automated backup as well as restoration functionalities to augment fault tolerance. Frequent backups facilitate the restoration of a database to a prior state in the occurrence of corrupted data or inadvertent deletions (Abadi, 2009).

7.2.3. Performance Optimization

The enhancement of performance becomes of utmost importance in order to achieve optimal effectiveness and responsiveness of cloud-based databases. Two crucial aspects to contemplate are the optimization of queries and the implementation of indexing strategies.

The process of query optimization entails the examination and refinement of the execution plans for database queries with the aim of enhancing their efficiency. The optimization of queries holds significant importance in cloud databases, primarily due to the possibility of distributed information as well as the imperative to reduce the influence of network traffic (Chen et al., 2022).

Cloud-based database platforms generally offer query optimization functionalities and tools that aid in enhancing query performance. The aforementioned characteristics may comprise query analyzers that detect impediments and propose enhancements and also indexing suggestions that are grounded on query patterns.

The implementation of indexing is crucial in enhancing the efficiency of query processing, as it facilitates expedited data retrieval. Cloud-based databases facilitate diverse indexing techniques, including B-tree, hash, or just bitmap indexes, contingent upon the data designs as well as query patterns (Zhang et al., 2021).

The process of indexing necessitates meticulous consideration of the workload along with the attributes of the data that is being retained. The act of over-indexing may have a negative impact on the performance of writing, whereas under-indexing may result in slower response times for queries. Consistent observation and adjustment of indexes are imperative for preserving peak efficiency as data quantities and inquiry patterns progress (Zhang et al., 2022).

7.2.4. Security and Compliance

The incorporation of measures to safeguard data and guarantee adherence to regulatory requirements are fundamental factors to contemplate in the development of cloud-based databases. Two crucial factors that require attention are the implementation of data encryption as well as the establishment of access control along with authentication mechanisms. The implementation of strong encryption methods is imperative for cloud databases to safeguard confidential information during transmission and storage. It is recommended to employ encryption protocols such as Transport Layer Security (TLS) or just Secure Sockets Layer (SSL) to ensure the security of data that is transmitted among clients as well as the database (Al-Aqrabi et al., 2012) (Figure 7.4).

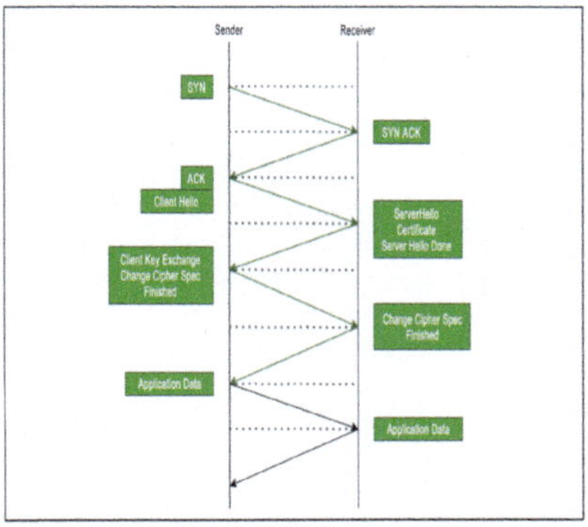

Figure 7.4. Diagram of transport layer security (TLS).

Source: Om Mishra, Creative Commons License.

It is recommended to apply data encryption while data is at rest in order to ensure the protection of data that is stored within a database. CSP frequently provide encoding mechanisms, including Transparent Information Encryption (TDE) or just Key Management Services (KMS), for secure data encoding as well as encryption key administration.

Cloud databases are required to implement appropriate access control as well as authentication measures to deter unauthorized access to confidential information. The task at hand pertains to the deployment of robust user authentication techniques, including multi-factor authentication (MFA) or just username/password, alongside roles-based access control (RBAC). RBAC enables precise regulation of database access through the allocation of distinct privileges and authorizations to individual users or groups of users. Furthermore, the implementation of audit logs is recommended to oversee and scrutinize database operations, guaranteeing responsibility and adherence to data protection statutes. Integrating access and authentication services are frequently provided by cloud providers, like IAM, which is provided by AWS, or Microsoft's Active Directory, to enable efficient management of individual logins and permissions (Doelitzscher et al., 2010; Bracci et al., 2012).

7.3. DATABASE-AS-A-SERVICE (DBAAS) PROVIDERS

Remember

When choosing a DBaaS provider, consider factors such as pricing, scalability, security features, and customer support.

7.3.1. Major Cloud Service Providers Offering DBaaS

7.3.1.1. Amazon Web Services (AWS)

AWS acts as a prominent provider of online services, offering a wide array of DBaaS solutions. AWS offers a variety of managed databases, each tailored to meet distinct data management needs.

a. Amazon's RDS offers a managed service for relational databases that provides support for widely used database engines, including SQL Server, MySQL, the case of Oracle, as well as SQL Server from Microsoft. The utilization of cloud technology streamlines the procedures involved in configuring, managing, and expanding a relational database. RDS enables users to concentrate on their apps without being concerned about the administration of infrastructure tasks (Difallah et al., 2013).

b. Amazon DynamoDB serves as a database without SQL service that is fully managed and offered by AWS. The system provides a high level of scalability, automated data replication, and optimal performance with minimal latency. DynamoDB has been engineered to cater to the demands of high-volume applications and offers a range of functionalities such as self-adjusting capacity, in-memory caching via DynamoDB Accelerator (DAX), as well as worldwide data replication through Global Tables.

c. Amazon Document DB has become a document database assistance that is fully managed and compatible with MongoDB. The system offers the advantageous traits language SQL database with regard to adaptability and capacity, while simultaneously ensuring compatibility with pre-existing MongoDB applications. The Document DB platform is specifically engineered to accommodate the storage, retrieval, and organization of JSON-based documents on a large scale (Gourisaria et al., 2020) (Figure 7.5).

Figure 7.5. Image of logo of MongoDB.

Source: Navindu Jayatilake, Creative Commons License.

d. Amazon Neptune seems to be a graph database assistance that is designed specifically for the storage and retrieval of interconnected data. It is an entirely managed offering. The software provides backing for the widely used graph query language known as Apache Tinker Pop Gremlin, as well as the Resource Description Framework (RDF) query language, commonly referred to as SPARQL. Neptune is specifically designed to enhance the efficiency and effectiveness of graph workloads, thereby providing superior levels of performance as well as scalability (Parida et al., 2019).

7.3.1.2. Microsoft Azure

Microsoft Azure has emerged as a prominent contender in the realm of cloud computing, providing an extensive selection of DBaaS solutions (Jain et al., 2017). Azure offers managed database resources that cater to both relational as well as NoSQL databases.

a. The Microsoft Azure SQL Database has become a relational database service that is completely managed, intelligent, as well as scalable. The software provides compatibility using SQL Server from Microsoft and incorporates functionalities such as automated tuning, inherent intelligence, and robust availability. The Azure SQL Database has been specifically developed to cater to high-priority workloads and offers a range of choices such as single databases or just elastic pools to effectively manage multiple databases (Mehak et al., 2014; Taft et al., 2016).
b. Azure Cosmos DB seems to be an online database that is distributed worldwide and supports various NoSQL data models, which include documents, key-value, charts, as well as column families. The aforementioned system provides a reliable assurance of minimal delay, automated organization of data, and worldwide duplication of information. The Cosmos DB system has been specifically developed to facilitate the construction of applications that demand both high levels of responsiveness and scalability, while also necessitating low-latency transmission of data from any location across the globe.
c. Azure offers managed databases over MySQL as well as PostgreSQL, providing features such as routine backups, high reliability, and automatic adjusting capabilities. The aforementioned services facilitate the effortless deployment, administration, and expansion of both MySQL as well as PostgreSQL relational databases within the Azure cloud milieu (Hans et al., 2013).

d. Azure Database over MariaDB is an online database that is completely managed, scalable, as well as secure. It is designed to be compatible with MariaDB. The software offers functionalities such as automated data backups, robust system availability, and sophisticated optimization of system performance (Bonthu et al., 2014).

7.3.1.3. Google Cloud Platform (GCP)

The Cloud Platform offered by Google (GCP) provides a variety of DBaaS solutions that are tailored to meet diverse database needs. The managed databases offered by GCP have been designed with a focus on achieving high availability, scalability, and reliability.

a. Cloud SQL offers a comprehensive managed service for relational databases offering support for both MySQL and PostgreSQL. The service provides automated backup functionality, and automatic scaling capabilities, and ensures high availability. Cloud SQL offers a database environment that is recognizable to users and can be effortlessly integrated with various GCP services (Shashi, 2023).

b. Cloud Spanner has become a relational database assistance that is horizontally scalable as well as distributed worldwide. This technology amalgamates the advantages of relational database systems with the expandability of NoSQL databases. Spanner offers robust consistency, elevated availability, and automated replication of information across numerous regions.

c. Cloud Fire store serves as a NoSQL document-oriented database system that provides instantaneous synchronization as well as automated scaling capabilities (Chi et al., 2023). The system is specifically engineered to facilitate the development of applications that can be scaled up and are responsive, while also enabling immediate updates as well as offline data access. Fire store offers functionalities such as automated searching, and querying, as well as security regulations.

d. Cloud Bigtable represents a NoSQL wide-column records service that can be fully managed, offering high scalability and low-latency retrieval of vast amounts of data (Gessert et al., 2016; Singh, 2016). This solution is well-suited for scenarios that necessitate a high volume of data processing and minimal delay, like time-series data, IoT applications, and analytical workloads.

7.3.2. Comparison of DBaaS Providers

7.3.2.1. Pricing Models

a. AWS provides diverse pricing structures for its DBaaS offerings. The pricing structure for Amazon RDS is determined by various factors, including the database engine, illustration type, storage capacity, and data transfer. The pricing structure for DynamoDB is determined by either the provisioned power mode or the on-demand power mode. The pricing structure for Amazon DocumentDB is determined by the amount of storage utilized and the size of the instance. The pricing structure for Amazon Neptune is determined by the size that comprises the illustration and the quantity of storage utilized (Curino et al., 2012).

b. Microsoft Azure provides various pricing models associated with its DBaaS offerings. The pricing structure for the Microsoft Azure SQL Database is determined by various factors, including the service tier, degree of performance, as well as the dimension of the database. The pricing structure for Azure Cosmos DB is determined by various factors, including but not limited to throughput, storage, as well as information transfer. The pricing structure for Azure Database offerings for the databases MySQL and PostgreSQL will be determined by each service tier as well as the vCore system. The pricing structure for Azure Database over MariaDB is determined by the chosen service tier as well as the vCore model (Calderoni et al., 2022).

c. GCP adheres to a pricing strategy that entails payment based on usage for its DBaaS offerings. The pricing structure for Cloud SQL is determined by various factors, including the type of instance, the amount of storage allocated, and the extent of network usage. The pricing model for Cloud Spanner is determined by the number of nodes, storage utilization, and network egress. The pricing structure for Cloud Firestore is determined by the number of operations performed, the amount of storage utilized, and the extent of network usage. The pricing structure for Cloud Bigtable is determined by various factors, including the number of nodes, capacity for storage, as well as network egress (Xiong et al., 2014; Astrova et al., 2017).

7.3.2.2. Features and Limitations

a. The AWS DBaaS services provide a range of functionalities including automated backups, scalability, enhanced availability, as well as read replicas to optimize performance. Notwithstanding, certain constraints consist of restricted authority over fundamental infrastructure and plausible vendor lock-in.

b. The Azure DBaaS services offered by Microsoft Azure encompass a range of functionalities such as automated tuning, embedded intelligence, worldwide data replication, and compatibility with diverse data models (Sharma, 2016). The constraints of the system may comprise possible impediments to optimal performance and restricted opportunities for customization.

c. The GCP provides DBaaS offerings that encompass a range of capabilities, including horizontal scalability, worldwide information replication, immediate updates, as well as low-latency access. Notwithstanding, constraints arise from the need to acquire expertise in certain services and probable intricacy in overseeing deployments across multiple regions (Willging et al., 2016).

To sum up, prominent cloud computing service providers, including AWS, Azure from Microsoft, as well as GCP, present a diverse array of DBaaS options (Haber et al., 2022). Every provider offers a distinct range of managed databases that come with varying pricing structures, functionalities, and constraints. The selection of an appropriate DBaaS provider is contingent upon various factors, including but not limited to, the database prerequisites, the extent of scalability necessities, the financial constraints, and the inclination towards a specific cloud service provider.

7.4. MIGRATING TO CLOUD DATABASES AND DBAAS

Tip

Implementing data migration strategies like parallel loading and change data capture can help streamline the migration process.

7.4.1. Evaluating the Migration Readiness

Prior to initiating the transition process towards cloud-based databases as well as DBaaS, it's imperative to assess the preparedness of the current database infrastructure (Pavlo et al., 2021). This assessment facilitates the identification of prospective obstacles and hazards that could emerge throughout the migration procedure. Several crucial factors should be taken into account during the process of evaluation, including:

a. Evaluation of the construction of the present on-premises or conventional database system. It is imperative to comprehend the database classification, namely interpersonal, NoSQL, or just NewSQL, as well as the intricacy of the schema, the magnitude of the data, and any potential modifications or amalgamations that may be present.
b. The speed and capacity specifications for the database should be assessed to determine the level of performance expectations as well as the scalability needed. It is important to take into account multiple variables such as the number of simultaneous users, the volume of transactions, and the patterns of data growth. This assessment aids in ascertaining whether the cloud-based database solution can fulfill the intended performance benchmarks (Wilging et al., 2016).
c. The task at hand involves conducting a thorough evaluation of the current security protocols and regulatory obligations pertaining to data protection. It is imperative to identify and take into account data governance regulations, restrictions on access, encryption processes, and regulatory frameworks throughout the migration process. It is imperative to verify that the selected cloud-based database solution offers sufficient security measures and certifications for compliance (Chandra et al., 2012).
d. Conduct a comprehensive cost analysis that contrasts the expenditures linked with the upkeep of the existing on-premises records systems with the financial implications of transferring to a cloud-based database or just DBaaS. It is imperative to consider various factors like hardware, software permits, maintenance, as well as ongoing operational expenses (Avram, 2014; Charette, 2005). This analysis serves to substantiate the financial advantages of transitioning to cloud computing.

7.4.2. Planning the Migration Process

Upon completion of the migration willingness evaluation, it is imperative to establish a clearly defined plan to facilitate a seamless and prosperous migration (Morgado et al., 2021). The subsequent procedures delineate the process of planning for migration.

a. It is imperative to establish a clear definition of the objectives as well as the objectives for the migration process. Identify the underlying factors that drive the migration process, including but not limited to financial benefits, ability to accommodate growth, and enhanced operational efficiency. The objectives shall serve as the impetus for the migration strategy as well as guarantee a concentrated approach.

b. Selecting an appropriate cloud database or DBaaS provider is crucial. Conduct an assessment of various CSPs that provide Cloud Databases and DBaaS solutions. It is advisable to take into account various factors such as the type of database backing, scalability alternatives, security features, cost structures, along with vendor reputation. Select a service provider that is in accordance with the particular needs and demands of the organization (Smith & Kumar, 2004; Papazoglou et al., 2006; Khan et al., 2023; Jamshidi et al., 2017).

c. Conducting a comprehensive data inventory is imperative to identify every record as well as information sets that require migration and subsequent classification. Categorize the information according to its level of importance, vulnerability to unauthorized access, and adherence to legal requirements. The aforementioned categorization shall aid in the prioritization of the migration pattern and the identification of the necessary security protocols (Al Shehri, 2013).

d. The Migration Approach is the process of determining and choosing the most suitable migration approach that aligns with the particular needs and limitations of the organization. Typical migration methodologies encompass a gradual progression, a trial migration, or a full-scale transition. Assess the requirement for a hybrid cloud strategy, wherein a subset of data may be stored in a cloud environment while retaining a segment of it on-premises.

e. Establishing a specialized migration team consisting of database managers, cloud designers, network professionals, as well as security experts is recommended in order to facilitate the migration process. Allocate distinct duties and obligations to the members of the team, guaranteeing efficient synchronization and correspondence during the entirety of the migration procedure (Cristian, 1991; Sakr, 2014; Kowalczyk et al., 2022)

f. In order to ensure a successful migration, it is imperative to establish a comprehensive migration timeline that outlines specific milestones, inspection points, as well as deadlines. It is imperative to take into account dependencies, constraints related to downtime, as well as potential risks when engaging in the procedure of planning. Effectively convey the chronological sequence of events to relevant parties and secure their agreement and endorsement (Alam & Shakil, 2013).

7.4.3. Data Migration Strategies

The transfer of data from the current database systems to the IoT is a crucial component of the migration procedure. There are two prevalent techniques for transferring data in cloud-based databases as well as DBaaS environments.

a. The Lift-and-Shift approach involves transferring the complete database to the cloud infrastructure without any substantial modifications to the database architecture or application code. The customary approach for transferring data involves the utilization of data restoration and backup methodologies or database export/import tools. The lift-and-shift methodology is deemed appropriate in situations where temporal limitations or just resource constraints hinder significant alterations to the database (Kanojia & Tanwani, 2022).

b. The proposed approach entails establishing a replication process between the local database as well as the cloud-based database to achieve database replication and synchronization. Modifications performed on the database located on the local premises are consistently duplicated and harmonized with the cloud-based database until a state of synchronization is achieved between both systems. Upon the completion of synchronizations, app traffic is rerouted toward the cloud database, along with the on-premises file is decommissioned (Guillonneau & Vallancien, 2000; Ezaki et al., 2004).

7.4.4. Best Practices for Successful Migration

In order to guarantee a prosperous transition to cloud relational databases as well as DBaaS, it is advisable to take into account the following optimal procedures:

a. Conduct data preprocessing by cleansing and optimizing the data prior to migration. The process entails eliminating redundant entries, addressing discrepancies, and guaranteeing the integrity of the data. The process of data preprocessing facilitates the efficient transfer of data and guarantees the preservation of data integrity during migration.

b. Undertake a pilot migration to assess the migration method and verify the outcomes by executing a pilot relocation with a limited subset of data. This facilitates the identification and resolution of any potential issues or challenges prior to the complete migration of the dataset (Hennebry & Preibisch, 2012).

c. Establishing mechanisms for data preservation and retrieval is crucial. Establishing backup and restoration of data protocols in the cloud is recommended to mitigate the risk of losing or corrupting information that may arise throughout the migration procedure. It is imperative to conduct routine assessments of backup and recovery protocols to ascertain their dependability and efficacy (Fjeldstad et al., 2018).

d. To enhance performance, it is recommended to leverage cloud-based features and optimizations within the cloud environment. Optimize query performance by configuring suitable indexes, partitioning tables as required, and refining query optimization. It is imperative to oversee the operational efficiency of the cloud-based database system throughout post-migration and undertake requisite modifications to guarantee optimal performance (Fehling et al., 2013).

e. It is imperative to consistently oversee the advancement of migration and keep a record of significant metrics, including transmission rate, latency, as well as error rates. Deploy suitable monitoring mechanisms and set up notifications to preemptively detect and address any potential problems that might emerge (Arora & Gupta, 2012).

f. It is recommended to conduct extensive testing subsequent to the migration process in order to verify the proper functionality of the migrated databases. Conducting tests to ensure data integrity, evaluating the functionality of an application, and performing performance benchmarking are essential tasks in the realm of software testing. It is advisable to promptly attend to any identified issues in order to minimize potential disruptions to business operations (Hugo, 2009).

Through adherence to these optimal methodologies, entities can effectively reduce potential hazards, decrease periods of system inactivity, and guarantee a prosperous transition to cloud-based databases and DBaaS solutions. A meticulously designed and implemented migration procedure empowers enterprises to capitalize on the advantages of cloud-based technologies, including enhanced scalability, cost-effectiveness, and superior manageability, while simultaneously safeguarding the confidentiality and safety of their data (Deka, 2013).

7.5. MANAGING AND MONITORING CLOUD DATABASES

Did You Know?

Provisioning and configuration management tools provided by DBaaS providers can simplify the setup and management of cloud database.

7.5.1. Provisioning and Configuration Management

Efficient administration of cloud databases necessitates the incorporation of provisioning as well as handling configurations as crucial components. This section will delve into the fundamental factors and optimal methodologies pertaining to the provisioning and configuration of cloud-based databases (Ray, 2016).

The process of providing a database. Selecting the suitable service model as well as deployment option is of utmost importance while establishing a cloud database, contingent upon the particular necessities. Cloud computing services commonly provide a range of alternatives, including IaaS, Platform-as-a-Service (PaaS), as well as DBaaS. Each alternative presents distinct benefits and factors to take into account (Marinho et al., 2018).

In the context of IaaS, users are afforded greater autonomy over the underlying infrastructure, thereby enabling them to customize database software along with settings to align with their specific requirements. In the context of PaaS and DBaaS, the responsibility of managing the infrastructure that underlies the service is assumed by the provider, while the user's focus is primarily directed toward the management of the database along with its configuration. In the process of provisioning, it is imperative to take into account various factors including but not limited to the necessary amount of storage, performance specifications, security measures, and adherence to legal requirements. The optimum effectiveness and expenditure efficiency of a database instance can be ensured by meticulously selecting the appropriate size or tier (Rayes & Salam, 2022).

Effective configuration management is crucial for ensuring optimal efficiency and safety of the database subsequent to its provisioning. The process entails configuring and administering various aspects of the database, including but not limited to memory allocation, buffering size, cache circumstances, replication arrangements, as well as network settings (Aceto et al., 2013; Xavier et al., 2016; Hilprecht et al., 2020).

The configuration parameters of cloud databases are subject to variation based on the particular technology employed. It is advisable to adhere to the guidelines and optimal methodologies provided by the vendor with regard to the configuration of the database. Furthermore, the utilization of automation tools as well as the implementation of infrastructure-as-code methodologies can facilitate the optimization of the configuration administration procedure. Ansible, Puppet, as well as Terraform, are software tools that can be employed to specify and oversee the intended configurations of databases as code. This approach enables version control, replication, and streamlined deployment (Lange et al., 2012).

7.5.2. Performance Monitoring and Optimization

The efficient operation and attainment of performance objectives of cloud databases are contingent upon the criticality of performance monitoring as well as optimization. The following section will expound on diverse methodologies and instruments utilized in monitoring and enhancing the efficiency of cloud-based databases. The process of tracking the efficacy of a cloud database entails the gathering and examination of metrics pertaining to the use of resources, turnaround times of queries, throughput, and various other pertinent performance indicators. The majority of the suppliers of cloud services provide integrated recording and monitoring functionalities that offer valuable insights into the well-being and efficiency of the database. Establishing performance baselines as well as configuring alerts for important metrics is a crucial step in proactively detecting performance bottlenecks or just anomalies. This facilitates the identification of problems and prompt implementation of remedial measures (Ying et al., 2009).

Furthermore, advanced monitoring as well as visualizing of database efficiency metrics can be achieved through the utilization of independent monitoring instruments and structures that include Prometheus, Grafana, as well as New Relic.

Efficiency optimization techniques can be implemented to enhance the efficiency of a cloud-based database. These strategies may include various approaches:

a. The process of query optimization involves the analysis and optimization of database queries, which can lead to a significant improvement in system performance. The implementation of strategies such as index optimizing, query rewriting, as well as schema design improvements can effectively mitigate query execution durations and enhance the overall efficiency of the system (Alam et al., 2013).
b. The implementation of a caching layer, including Redis or Memcached, may successfully reduce the burden on the database by retaining frequently used information in memory. This approach can lead to an enhancement because of time.
c. Partitioning along with sharding are techniques used to distribute the computational load and improve scalability as well as efficiency by dividing large tables or just databases into multiple nodes or shards. It is important to give thoughtful consideration to both consistency of data and dissemination strategies.
d. Scaling databases either upward (scaling up) or just across (scaling out) has the potential to enhance efficiency. Vertical scaling pertains to the process of augmenting the resources, such as CPU and memory, for the instance that stores data. On the other hand, horizontal scaling pertains to the act of incorporating additional instance databases to facilitate the distribution of the workload (Tan et al., 2017).
e. Performance tuning involves the adjustment of diverse database arrangement parameters, including buffer sizes, conjunction pooling, as well as query analogy, with the aim of optimizing system performance. It is advisable to conduct benchmarking along with load testing procedures in order to detect any performance bottlenecks as well as subsequently maximize the configuration as deemed necessary.

Activity 7.1

In the context of cloud databases and Database-as-a-Service (DBaaS), how can organizations ensure the security and privacy of their data when migrating sensitive customer information to a cloud-based database?

7.6. SUMMARY

This chapter explores the fascinating world of cloud databases and the concept of DBaaS. It begins by tracing the evolution of DBaaS in the context of the ever-expanding cloud computing landscape. The chapter highlights the key factors that have propelled the popularity of DBaaS, including scalability, elasticity, availability, fault tolerance, performance optimization, and security.

The chapter provides valuable insights into the major providers of DBaaS, offering a comprehensive overview of the database options they offer. It sheds light on the wide range of databases available, including relational, NoSQL, and NewSQL databases, enabling readers to make informed decisions based on their specific requirements.

One of the fundamental aspects discussed in this chapter is the process of migrating to cloud databases and DBaaS. It emphasizes the importance of assessing the readiness of the current system before initiating the migration process. By planning meticulously, considering factors such as data migration strategies, and following best practices, organizations can ensure a smooth and successful migration to the cloud.

Managing and monitoring cloud databases is another critical aspect covered in this chapter. It highlights the significance of provisioning and configuration management, as well as the tools and services provided by DBaaS providers to simplify these tasks. The chapter also underscores the importance of performance monitoring and optimization, enabling organizations to maximize the efficiency and responsiveness of their cloud databases. In conclusion, Chapter 8 serves as a comprehensive guide to understanding cloud databases and DBaaS. It equips readers with the knowledge necessary to evaluate DBaaS providers, plan and execute seamless migrations, and effectively manage and monitor cloud databases. By embracing the power of DBaaS, organizations can leverage the scalability, flexibility, and security offered by the cloud to enhance their data management capabilities and drive innovation in their respective industries.

REVIEW QUESTION

1. What are the advantages of using Database-as-a-Service (DBaaS)?
2. Name three major cloud service providers offering DBaaS.
3. What are the key factors to consider when migrating to cloud databases?
4. Explain the concept of scalability and elasticity in the context of DBaaS.
5. How can data migration strategies like parallel loading and CDC facilitate the migration process?
6. What are the best practices for managing and monitoring cloud databases?
7. Why is performance monitoring important for cloud databases?
8. Discuss the security and compliance considerations when using DBaaS.
9. Compare and contrast relational, NoSQL, and NewSQL databases in the context of DBaaS.
10. How can provisioning and configuration management tools provided by DBaaS providers simplify database setup and management?

REFERENCES

1. Abadi, D. J., (2009). Data management in the cloud: Limitations and opportunities. *IEEE Data Eng. Bull., 32*(1), 3–12.
2. Abbasi, A., Sarker, S., & Chiang, R. H., (2016). Big data research in information systems: Toward an inclusive research agenda. *Journal of the Association for Information Systems, 17*(2), 3.
3. Aceto, G., Botta, A., De Donato, W., & Pescapè, A., (2013). Cloud monitoring: A survey. *Computer Networks, 57*(9), 2093–2115.
4. Achar, S., (2019). Cloud-based system design. *International Journal of All Research Education and Scientific Methods (IJARESM), 7*(8), 23–30.
5. Agrawal, D., El Abbadi, A., Das, S., & Elmore, A. J., (2011). Database scalability, elasticity, and autonomy in the cloud. In: *Database Systems for Advanced Applications: 16th International Conference, DASFAA 2011, Hong Kong, China, April 22–25, 2011, Proceedings, Part I 16* (pp. 2–15). Springer Berlin Heidelberg.
6. Al Shehri, W., (2013). Cloud database database as a service. *International Journal of Database Management Systems*, *5*(2), 1.
7. Alam, B., Doja, M. N., Mansaf, A., & Shweta, M., (2013). 5-layered architecture of cloud database management system. *AASRI Procedia, 5*, 194–199.
8. Alam, M., & Shakil, K. A., (2013). Cloud database management system architecture. *UACEE International Journal of Computer Science and its Applications*, *3*(1), 27–31.
9. Al-Aqrabi, H., Liu, L., Xu, J., Hill, R., Antonopoulos, N., & Zhan, Y., (2012). Investigation of IT security and compliance challenges in security-as-a-service for cloud computing. In: *2012 IEEE 15th International Symposium on Object/Component/Service-Oriented Real-Time Distributed Computing Workshops* (pp. 124–129). IEEE.
10. Al-Refai, M. N., & Haya, A., (2021). Database as a service (DBaaS) challenges and solutions. In: *2021 22nd International Arab Conference on Information Technology (ACIT)* (pp. 1–6). IEEE.
11. Arora, I., & Gupta, A., (2012). Cloud databases: A paradigm shift in databases. *International Journal of Computer Science Issues (IJCSI)*, *9*(4), 77.
12. Astrova, I., Koschel, A., Eickemeyer, C., Kersten, J., & Offel, N., (2017). DBaaS comparison: Amazon vs. Microsoft. In: *2017 International Conference on Information Society (i-Society)* (pp. 15–21). IEEE.

13. Avram, M. G., (2014). Advantages and challenges of adopting cloud computing from an enterprise perspective. *Procedia Technology, 12*, 529–534.
14. Bonthu, S., Thammiraju, S. D. M., & Murthy, Y. S. S. R., (2014). Study on database virtualization for database as a service (DBaaS). *International Journal of Advanced Research in Computer Science, 5*(2), 31–34.
15. Botta, A., De Donato, W., Persico, V., & Pescapé, A., (2016). Integration of cloud computing and internet of things: A survey. *Future Generation Computer Systems, 56*, 684–700.
16. Bracci, F., Corradi, A., & Foschini, L., (2012). Database security management for healthcare SaaS in the Amazon AWS cloud. In: *2012 IEEE Symposium on Computers and Communications (ISCC)* (pp. 000812–000819). IEEE.
17. Calderoni, L., Maio, D., & Tullini, L., (2022). Benchmarking cloud providers on serverless iot back-end infrastructures. *IEEE Internet of Things Journal, 9*(16), 15255–15269.
18. Chandra, D. G., Prakash, R., & Lamdharia, S., (2012). A study on cloud database. In: *2012 Fourth International Conference on Computational Intelligence and Communication Networks* (pp. 513–519). IEEE.
19. Charette, R. N., (2005). Why software fails [software failure]. *IEEE Spectrum, 42*(9), 42–49.
20. Chen, Z., Yang, X., Li, F., Cheng, X., Hu, Q., Miao, Z., & Wang, S., (2022). CloudJump: Optimizing cloud databases for cloud storages. *Proceedings of the VLDB Endowment, 15*(12), 3432–3444.
21. Chi, Y., Kemajou, V., Rajan, A., & Samuel, R., (2023). Cloud-based real-time well engineering: Coupling torque-and-drag and uncertainty modeling. In: *SPE/IADC Drilling Conference and Exhibition* (p. D021S014R005). SPE.
22. Cristian, F., (1991). Reaching agreement on processor-group membership in synchronous distributed systems. *Distributed Computing, 4*, 175–187.
23. Curino, C. A., Difallah, D. E., Pavlo, A., & Cudre-Mauroux, P., (2012). Benchmarking OLTP/web databases in the cloud: The OLTP-bench framework. In: *Proceedings of the Fourth International Workshop on Cloud Data Management* (pp. 17–20).

24. Deka, G. C., (2013). A survey of cloud database systems. *It Professional, 16*(2), 50–57.

25. Difallah, D. E., Pavlo, A., Curino, C., & Cudre-Mauroux, P., (2013). OLTP-bench: An extensible testbed for benchmarking relational databases. *Proceedings of the VLDB Endowment, 7*(4), 277–288.

26. Doelitzscher, F., Reich, C., & Sulistio, A., (2010). Designing cloud services adhering to government privacy laws. In: *2010 10th IEEE International Conference on Computer and Information Technology* (pp. 930–935). IEEE.

27. Ezaki, T., Kondo, K., Ozaki, H., Sasaki, N., Yonernura, H., Kitano, M., & Hirayarna, T., (2004). A 160Gb/s interface design configuration for multichip LSI. In: *2004 IEEE International Solid-State Circuits Conference (IEEE Cat. No. 04CH37519)* (pp. 140–141). IEEE.

28. Fehling, C., Leymann, F., Ruehl, S. T., Rudek, M., & Verclas, S., (2013). Service migration patterns–decision support and best practices for the migration of existing service-based applications to cloud environments. In: *2013 IEEE 6th International Conference on Service-Oriented Computing and Applications* (pp. 9–16). IEEE.

29. Ferretti, L., Colajanni, M., & Marchetti, M., (2013). Distributed, concurrent, and independent access to encrypted cloud databases. *IEEE Transactions on Parallel and Distributed Systems, 25*(2), 437–446.

30. Fjeldstad, H. P., Pulg, U., & Forseth, T., (2018). Safe two-way migration for salmonids and eel past hydropower structures in Europe: A review and recommendations for best-practice solutions. *Marine and Freshwater Research, 69*(12), 1834–1847.

31. Gessert, F., & Ritter, N., (2016). Scalable data management: NoSQL data stores in research and practice. In: *2016 IEEE 32nd International Conference on Data Engineering (ICDE)* (pp. 1420–1423). IEEE.

32. Gourisaria, M. K., Samanta, A., Saha, A., Patra, S. S., & Khilar, P. M., (2020). An extensive review on cloud computing. *Data Engineering and Communication Technology: Proceedings of 3rd ICDECT-2K19*, 53–78.

33. Guillonneau, B., & Vallancien, G., (2000). Laparoscopic radical prostatectomy: The Montsouris experience. *The Journal of Urology, 163*(2), 418–422.

34. Haber, M. J., Chappell, B., & Hills, C., (2022). Cloud service providers. In: *Cloud Attack Vectors: Building Effective Cyber-Defense Strategies to Protect Cloud Resources* (pp. 27–43). Berkeley, CA: Apress.

35. Hans, R., Dahlen, D., Zöller, S., Schuller, D., & Lampe, U., (2013). Enabling virtual manufacturing enterprises with cloud computing: An analysis of criteria for the selection of database as a service offers. In: *Advances in Sustainable and Competitive Manufacturing Systems: 23rd International Conference on Flexible Automation & Intelligent Manufacturing* (pp. 427–438). Springer International Publishing.
36. Helali, L., & Omri, M. N., (2021). A survey of data center consolidation in cloud computing systems. *Computer Science Review, 39*, 100366.
37. Hennebry, J. L., & Preibisch, K., (2012). A model for managed migration? Re - examining best practices in Canada's seasonal agricultural worker program. *International Migration, 50*, e19–e40.
38. Hilprecht, B., Binnig, C., & Röhm, U., (2020). Learning a partitioning advisor for cloud databases. In: *Proceedings of the 2020 ACM SIGMOD International Conference on Management of Data* (pp. 143–157).
39. Hugo, G., (2009). Best practice in temporary labor migration for development: A perspective from Asia and the Pacific. *International Migration, 47*(5), 23–74.
40. Jain, A., Mahajan, N., Jain, A., & Mahajan, N., (2017). Introduction to database as a service. *The Cloud DBA-Oracle: Managing Oracle Database in the Cloud,* 11–22.
41. Jamshidi, P., Pahl, C., & Mendonça, N. C., (2017). Pattern - based multi - cloud architecture migration. *Software: Practice and Experience, 47*(9), 1159–1184.
42. Kanojia, A., & Tanwani, S., (2022). IT infrastructure for smart city: Issues and challenges in migration from relational to NoSQL databases. *Decision Analytics for Sustainable Development in Smart Society 5.0: Issues, Challenges and Opportunities*, 93–112.
43. Khan, W., Kumar, T., Zhang, C., Raj, K., Roy, A. M., & Luo, B., (2023). SQL and NoSQL database software architecture performance analysis and assessments—A systematic literature review. *Big Data and Cognitive Computing, 7*(2), 97.
44. Kowalczyk, M., Marcinkowski, B., & Przybyłek, A., (2022). Scaled agile framework. Dealing with software process - related challenges of a financial group with the action research approach. *Journal of Software: Evolution and Process, 34*(6), e2455.
45. Lange, B., Koenig, S., McConnell, E., Chang, C. Y., Juang, R., Suma, E., & Rizzo, A., (2012). Interactive game-based rehabilitation using

the Microsoft Kinect. In: *Virtual Reality Conference, IEEE* (pp. 171–172). IEEE Computer Society.

46. Lehner, W., & Sattler, K. U., (2010). Database as a service (DBaaS). In*: 2010 IEEE 26th International Conference on Data Engineering (ICDE 2010)* (pp. 1216–1217). IEEE.
47. Li, G., Dong, H., & Zhang, C., (2022). Cloud databases: New techniques, challenges, and opportunities. *Proceedings of the VLDB Endowment, 15*(12), 3758–3761.
48. Marinho, C. S., Moreira, L. O., Coutinho, E. F., Costa, F. J. S., Sousa, F. R., & Machado, J. C., (2018). LABAREDA: A predictive and elastic load balancing service for cloud-replicated databases. *Journal of Information and Data Management, 9*(1), 94.
49. Mehak, F., Masood, R., Ghazi, Y., Shibli, M. A., & Khan, S., (2014). Security aspects of database-as-a-service (DBaaS) in cloud computing. *Cloud Computing: Challenges, Limitations and R&D Solutions*, 297–324.
50. Morgado, J. C., Lencastre, J. A., Freires, T., & Bento, M., (2021). Smart education as empowerment: Outlining veteran teachers' training to promote digital migration. *Technology, Knowledge, and Learning, 26*(4), 897–916.
51. Munir, K., (2015). Security model for cloud database as a service (DBaaS). In: *2015 International Conference on Cloud Technologies and Applications (CloudTech)* (pp. 1–5). IEEE.
52. Papazoglou, M. P., & Van, D. H. W. J., (2006). Service-oriented design and development methodology. *International Journal of Web Engineering and Technology, 2*(4), 412–442.
53. Parida, T., Pal, S., & Sarkar, A., (2019). SaaS level database integration in cloud environment through database as a service. *International Journal of Services Technology and Management, 25*(3, 4), 236–266.
54. Pavlo, A., Butrovich, M., Ma, L., Menon, P., Lim, W. S., Van, A. D., & Zhang, W., (2021). Make your database system dream of electric sheep: Towards self-driving operation. *Proceedings of the VLDB Endowment, 14*(12), 3211–3221.
55. Radhakrishnan, G., (2012). Adaptive application scaling for improving fault-tolerance and availability in the cloud. *Bell Labs Technical Journal, 17*(2), 5–14.

56. Ray, P. P., (2016). A survey of IoT cloud platforms. *Future Computing and Informatics Journal, 1*(1, 2), 35–46.
57. Rayes, A., & Salam, S., (2022). IoT services platform: Functions and requirements. In: *Internet of Things from Hype to Reality: The Road to Digitization* (pp. 179–211). Cham: Springer International Publishing.
58. Sakr, S., (2014). Cloud-hosted databases: Technologies, challenges, and opportunities. *Cluster Computing, 17*, 487–502.
59. Shaik, B., Vallarapu, A., Shaik, B., & Vallarapu, A., (2018). Introduction to databases in the cloud. *Beginning PostgreSQL on the Cloud: Simplifying Database as a Service on Cloud Platforms*, 1–31.
60. Sharma, S., (2016). Expanded cloud plumes hiding big data ecosystem. *Future Generation Computer Systems, 59*, 63–92.
61. Shashi, A., (2023). Google cloud SQL. In: *Designing Applications for Google Cloud Platform: Create and Deploy Applications Using Java* (pp. 119–153). Berkeley, CA: Apress
62. Singh, A., (2016). NoSQL: A new horizon in big data. *International Journal of Scientific Research in Science, Engineering and Technology, 2*(2), 1183–1188.
63. Smith, M. A., & Kumar, R. L., (2004). A theory of application service provider (ASP) use from a client perspective. *Information & Management, 41*(8), 977–1002.
64. Taft, R., Lang, W., Duggan, J., Elmore, A. J., Stonebraker, M., & DeWitt, D., (2016). STeP: Scalable tenant placement for managing database-as-a-service deployments. In: *Proceedings of the Seventh ACM Symposium on Cloud Computing* (pp. 388–400).
65. Tan, D. P., Li, L., Zhu, Y. L., Zheng, S., Ruan, H. J., & Jiang, X. Y., (2017). An embedded cloud database service method for distributed industry monitoring. *IEEE Transactions on Industrial Informatics, 14*(7), 2881–2893.
66. Verbitski, A., Gupta, A., Saha, D., Brahmadesam, M., Gupta, K., Mittal, R., & Bao, X., (2017). Amazon aurora: Design considerations for high throughput cloud-native relational databases. In: *Proceedings of the 2017 ACM International Conference on Management of Data* (pp. 1041–1052).
67. Vidhyalakshmi, R., & Kumar, V., (2017). CORE framework for evaluating the reliability of SaaS products. *Future Generation Computer Systems, 72*, 23–36.

68. Willging, C. E., Israel, T., Ley, D., Trott, E. M., DeMaria, C., Joplin, A., & Smiley, V., (2016). Coaching mental health peer advocates for rural LGBTQ people. *Journal of Gay & Lesbian Mental Health, 20*(3), 214–236.

69. Xavier, M. G., Matteussi, K. J., Lorenzo, F., & De Rose, C. A., (2016). Understanding performance interference in multi-tenant cloud databases and web applications. In: *2016 IEEE International Conference on Big Data (Big Data)* (pp. 2847–2852). IEEE.

70. Xiong, P., Chi, Y., Zhu, S., Moon, H. J., Pu, C., & Hacgümüş, H., (2014). SmartSLA: Cost-sensitive management of virtualized resources for CPU-bound database services. *IEEE Transactions on Parallel and Distributed Systems, 26*(5), 1441–1451.

71. Ying, Z., & Yong, S., (2009). Cloud storage management technology. In: *2009 Second International Conference on Information and Computing Science* (Vol. 1, pp. 309–311). IEEE.

72. Zhang, J., Zhou, K., Li, G., Liu, Y., Xie, M., Cheng, B., & Xing, J., (2021). CDBTune+: An efficient deep reinforcement learning-based automatic cloud database tuning system. *The VLDB Journal, 30*(6), 959–987.

73. Zhang, X., Wu, H., Li, Y., Tan, J., Li, F., & Cui, B., (2022). Towards dynamic and safe configuration tuning for cloud databases. In: *Proceedings of the 2022 International Conference on Management of Data* (pp. 631–645).

74. Zhang, Y., Ruan, C., Li, C., Yang, X., Cao, W., Li, F., & Bi, C., (2021). Towards cost-effective and elastic cloud database deployment via memory disaggregation. *Proceedings of the VLDB Endowment, 14*(10), 1900–1912.

75. Zukin, S., & Braslow, L., (2011). The life cycle of New York's creative districts: Reflections on the unanticipated consequences of unplanned cultural zones. *City, Culture and Society, 2*(3), 131–140.

CHAPTER

8

Distributed Database Systems

CONTENT

8.1. INTRODUCTION

A distributed database system comprises multiple computers, each of which stores its own copy of the database. These computers communicate with one another, typically using high-speed networks or telephone lines, in order to share information. It is not unheard of for these various devices to have diverse technical specifications due to their relative relevance and location in the system. It is important to consider that the client-server systems that served as the basis for our earlier discussion in this chapter are rather different from the distributed database systems described here. In client-server architectures, the data is not itself distributed; rather, it is kept on a server machine, which can be retrieved in a decentralized manner by client machines. On the other hand, when using a distributed database system, the data is spread out over a variety of computers rather than being kept on a single server. Because of this, decisions need to be made regarding how the data are going to be distributed and also regarding how updates are going to be propagated over several different sites (Ozsu & Valduriez, 1991).

The distribution of data by an organization throughout its many locations and departments enables data to be stored where it was generated or, more accurately, where it is used the most while making it simple to retrieve the data from other locations. The following diagram illustrates a distributed system's basic organizational framework (Bernstein & Goodman, 1981) (Figure 8.1):

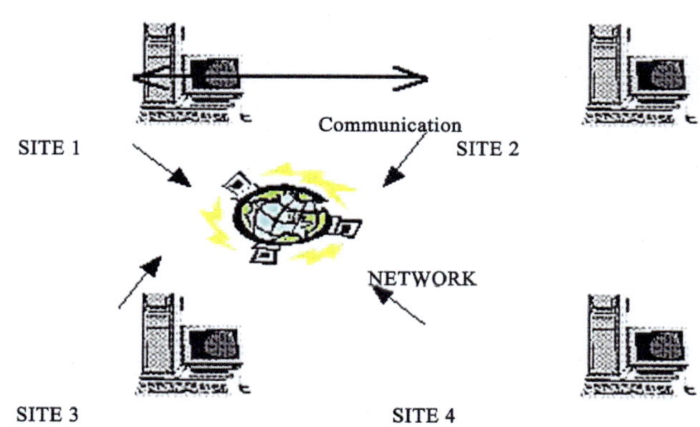

Figure 8.1. Illustration of diagram illustrates a distributed system's basic organizational framework.

Source: CS UCT, Creative Commons License.

The example illustrates that each distinct site essentially functions as an independent database system site, with each location possessing its own databases and also its own transaction management software and database management system (DBMS) (Mullin, 1990). In the discourse surrounding distributed database systems, a prevalent assumption is that the multiple databases implicated are geographically dispersed from one another. It is crucial to note that the geographical separation of databases has minimal impact on the functioning of the distributed system. The technical challenges that arise when databases are logically separated are similar to those encountered when a considerable distance separates them. What are the reasons behind the increasing popularity of distributed databases? There exist several reasons that advocate for the adoption of a distributed database system, including (Thomas et al., 1990).

1. The sharing of data;
2. Local autonomy;
3. Data availability;
4. Improving performance; an
5. Improving system reliability.

Moreover, probably, the entity that opted to integrate the system will also be dispersed. This implies that a company's organizational structure typically comprises multiple departments and divisions. Utilizing distributed database systems can yield advantageous outcomes, which can be better comprehended by incorporating an illustrative instance (Traiger et al., 1982).

8.2. SCENARIO BANKING SYSTEM

Consider a hypothetical banking system that functions across multiple distinct locations. For this illustration, we shall examine two branches in Manchester and Birmingham, respectively. The account data pertaining to Manchester accounts are kept within the geographical boundaries of Manchester, whereas the account data of Birmingham is stored within the confines of Birmingham. The utilization of this system provides two significant advantages. Firstly, it enhances the processing efficiency by storing the data in the location where it is often accessed. Secondly, it improves the accessibility of account data (Raja, 2012).

Despite its advantages, utilizing distributed database systems is not exempt from certain limitations. The primary drawback pertains to the increased intricacy associated with ensuring adequate coordination among multiple sites. The augmentation in intricacy can manifest in diverse manners (Singh, 2012):

1. Multiple databases operating simultaneously pose a challenge in ensuring the accuracy of the system's operational algorithms, thereby increasing the likelihood of encountering bugs. There exists a possibility for the presence of highly nuanced software defects.
2. The augmented processing overhead resulting from the need to attain inter-site coordination constitutes a significant burden that is absent in centralized systems (Ahmad et al., 2021).

According to Date (1999), the fundamental principle of a truly distributed database is that it should appear to the user as a non-distributed system.

To achieve this fundamental principle, twelve subordinate regulations have been instituted. The following twelve objectives are delineated (Hu et al., 2012) (Figure 8.2):

1. Local autonomy;
2. No reliance on a central site;
3. Continuous operation;
4. Operating system independence;
5. Distributed transaction management;
6. Location independence;

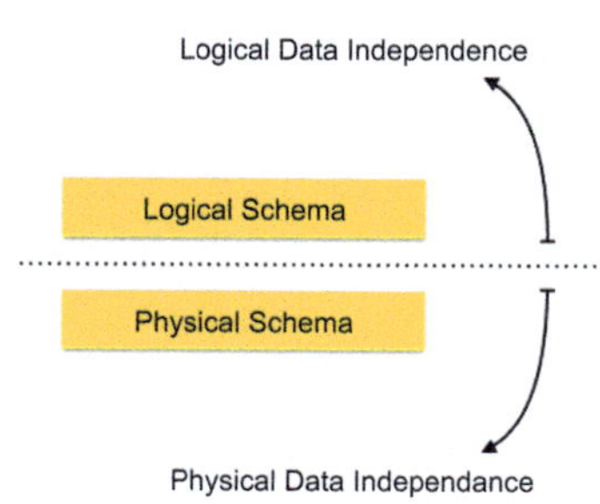

Figure 8.2. Schematic of type of DBMS type (data independence).

Source: Tutorialspoint, Creative Commons License.

7. Fragmentation independence;
8. Replication independence;

9. Distributed query processing;
10. Hardware Independence;
11. Network independence; and
12. DBMS independence.

The interdependence of these twelve objectives is evident. Additionally, it should be noted that these findings may not be comprehensive and do not hold equal weight in terms of significance.

The primary concern in distributed database systems pertains to the propagation of updates across the system. Two pivotal concepts heavily influence this process (Gupta et al., 2017) (Figure 8.3).

1. *Data fragmentation* refers to dividing the components of a comprehensive database into distinct locations.
2. *Data replication* refers to the procedure of ensuring that modifications made to data are consistently reflected across multiple locations (Nayak & Chandiramani, 2022).

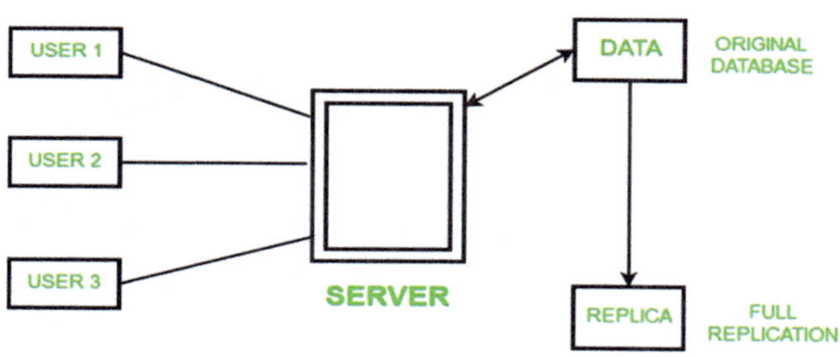

Figure 8.3. Image of data replication process.

Source: Harshita Pandey, Creative Commons License.

8.3. FRAGMENTATION INDEPENDENCE

The notion of fragmentation independence holds significant importance within distributed database systems. The term pertains to the capacity of a dispersed database to undergo fragmentation or segmentation into smaller units while preserving the distinctiveness and self-governance of each segment. Fragmentation independence refers to the capability of each fragment within a distributed database to function autonomously without relying on other fragments or the system as a whole (Foster et al., 2016).

Remember

Fragmentation independence enables efficient data distribution and replication strategies in distributed database systems.

Fragmentation is a process in which a distributed database is partitioned into smaller units known as fragments, which can be allocated across various nodes or sites within the distributed system. Fragmentation can be achieved using various criteria, including horizontal fragmentation, which involves dividing the database vertically based on rows, and vertical fragmentation, which involves dividing the database horizontally centered on columns. Fragmentation independence guarantees that every fragment functions as an autonomous entity with distinct operations and functionalities, irrespective of the fragmentation method employed (Sacco, 1986) (Figure 8.4).

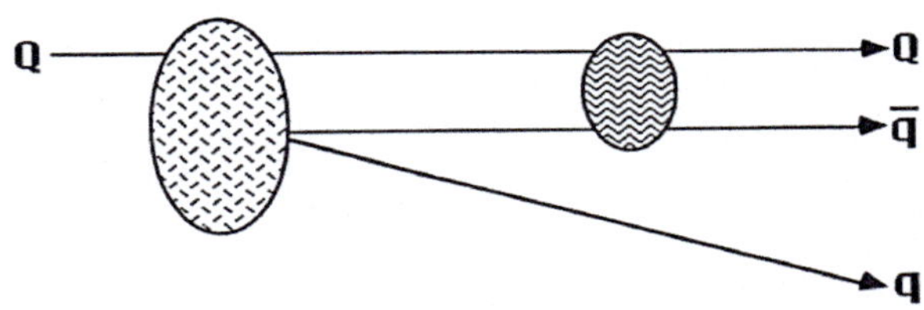

Figure 8.4. *Image of diagram of independent fragmentation.*

Source: David R, Creative Commons License.

The utilization of fragmentation independence in a distributed database system presents various benefits.

1. **Autonomy:** The concept of autonomy is manifested in the individuality of each fragment, which possesses its own distinct schema, data, and processing capabilities. This feature enables individual fragments to function autonomously without being dependent on other pieces or the central system. The system's performance and availability are improved by the implementation of autonomy, whereby each fragment can independently manage its own transactions and inquiries within its local environment (Ulus & Uysal, 2003).
2. **Scalability***:* The property of fragmentation independence allows for horizontal scalability in a distributed database system by

adding additional fragments to the system. With fresh fragments, the system can distribute the workload among them, enabling it to manage a greater volume of data and transactions. The ability to scale is of utmost importance to address the growing data requirements and enhance the system's efficiency.

3. **Fault Tolerance:** The enhancement of fault tolerance is achieved through the distribution of the database across multiple fragments, as facilitated by fragmentation independence. In case of a fragment failure or network outage, the remaining fragments can operate autonomously, thereby preserving the system's functionality. Implementing fault tolerance mechanisms mitigates the potential for a singular point of failure, thereby augmenting the overall dependability of the distributed database system (Kumar et al., 2013).
4. **Localization of Data:** The property of fragmentation independence enables the localization or colocation of data based on specific criteria. Data pertaining to a specific geographic area or organizational division can be consolidated into fragments, resulting in enhanced data retrieval and decreased network latency. Fragmentation independence improves retrieval of data performance and enhances the overall user experience by ensuring that data is located close to the users or applications that need it.

In order to attain fragmentation independence, distributed database systems frequently utilize strategies such as data replication, data partitioning, and distributed query processing. The techniques above guarantee the self-sufficient and effective functioning of individual fragments while simultaneously upholding coherence, synchronization, and inter-fragment communication within the system.

Fragmentation independence is an essential component of distributed database systems. The fragmentation of a database into smaller units that can function independently provides several advantages, including but not limited to autonomy, scalability, fault tolerance, and localized data. Through the utilization of fragmentation independence, entities have the ability to create and implement resilient distributed database systems that possess the capacity to manage extensive data processing and furnish dependable services within distributed computing settings (Grebla & Cenan, 2005).

8.4. REPLICATION INDEPENDENCE

Replication independence, in the context of a distributed database system, pertains to the system's capacity to manage data replication without compromising the accuracy and coherence of the data. The transparent occurrence of the replication process is guaranteed while maintaining the general integrity of the database.

The concept of replication is a fundamental aspect in distributed databases, wherein data is duplicated and saved on numerous nodes or servers to enhance performance, fault tolerance, and availability. Ensuring consistency among replicated copies can pose a challenge, particularly when faced with concurrent updates and network latency. The issue of managing data replication effectively is tackled by implementing mechanisms that ensure replication independence.

In order to attain replication independence, it is necessary to take into account various crucial factors.

1. **Consistency Models:** Consistency models refer to a set of guidelines and assurances that dictate the maintenance of consistency in replicated data. Various consistency models, namely strong consistency, eventual consistency, and causal consistency, offer different levels of replication autonomy. In the realm of distributed systems, strong consistency models are designed to maintain uniformity across all replicas by ensuring that they consistently display the same value at all times. On the other hand, eventual consistency models permit temporary inconsistencies that are eventually resolved and lead to convergence. The selection of a consistency model is contingent upon the specific demands of the application and the balancing of trade-offs between performance and consistency (Bratsberg et al., 1997).
2. **Replication Protocols:** They govern the dissemination of modifications and updates among replicas. Diverse protocols, namely primary-copy, multi-primary, and quorum-based, offer distinct methodologies for achieving replication autonomy. The aforementioned protocols are responsible for overseeing the replication factors, resolving conflicts, and guaranteeing the accurate application of updates to replicas, all while preserving the integrity of the data.

3. **Conflict Resolution**: In the context of data management, the occurrence of concurrent updates across multiple replicas can give rise to conflicts, which in turn can result in data inconsistencies. Effective conflict resolution strategies are, therefore, essential to ensure data integrity. The attainment of replication independence necessitates the implementation of efficient conflict resolution mechanisms that can effectively detect and resolve conflicts suitably. Various conflict resolution techniques, such as timestamp ordering, version vectors, distributed locks, and consensus protocols, such as Paxos or Raft, are available. These mechanisms guarantee the coordination and consistent application of conflicting updates among replicas.
4. **Synchronization and Communication:** The attainment of replication independence is contingent upon effectively implementing synchronization and communication protocols among the replicas. Various techniques, including replication logs, message passing, and gossip protocols, facilitate the distribution of updates and metadata among replicas. The process of synchronization guarantees that replicas remain current and maintain coherence with the most recent modifications, while communication protocols mitigate the effects of network latencies and malfunctions.
5. **Failure Handling:** The topic of concern is the management of failures. The issue of replication independence should consider the management of failures to uphold the coherence of data even when node failures or network partitions occur. The utilization of methods such as automatic failover, detection of replication lag, and resynchronization mechanisms aid in the recovery from system failures and the preservation of the coherence of duplicated data.

A distributed database system can attain replication autonomy through meticulous consideration of these factors. The implementation of this mechanism guarantees the ability to duplicate data across numerous nodes while maintaining the accuracy and uniformity of the database.

The property of replication independence facilitates the ability to scale, enhance fault tolerance, and optimize performance within distributed systems, all while preserving the integrity and dependability of data (Hellerstein et al., 1999).

8.5. UPDATE STRATEGIES FOR REPLICATED AND NON-REPLICATED DATA

Duplicating data in a distributed database system is a crucial factor in attaining elevated availability, resilience against faults, and enhanced efficiency. The replication process entails preserving numerous data duplicates throughout diverse nodes within the network, thereby enabling local accessibility and mitigating network latency. In the context of replicated data, it is imperative to exercise prudence in devising updated strategies that guarantee consistency and integrity among all replicas. The utilization of consensus protocols, such as the Paxos or Raft algorithms, is a frequently employed method. Implementing these protocols ensures the synchronized and coordinated application of updates, thereby ensuring the eventual attainment of a consistent state by all replicas. Furthermore, it is possible to ensure consistency by utilizing techniques such as primary-copy replication. This involves designating one replica as the primary and assigning it the responsibility of handling all updates (Du et al., 1993).

Tip

Choosing the appropriate update strategy is crucial for maintaining data consistency and ensuring reliable operations in distributed database systems.

In contrast, non-replicated data pertains to data that is stored and retrieved from a solitary location within the distributed database system. The update methods for non-replicated data are comparatively uncomplicated due to the absence of replicas. Local execution of non-replicated data updates can be performed on the node that hosts the data without necessitating collaboration with other nodes. The utilization of uncomplicated update strategies for non-replicated data presents benefits in the form of decreased overhead and enhanced performance. Nevertheless, it presents a potential vulnerability by creating a sole point of failure. In the event of a node failure that stores non-replicated data, the data becomes inaccessible until the node is either restored or replaced (Sivasubramanian et al., 2005).

The development of efficient, informed methods for replicated and non-replicated data in a distributed database system is of utmost importance to guarantee data consistency, availability, and reliability. The presence of replicated data requires the implementation of meticulous synchronization and coordination mechanisms to ensure uniformity among replicas. Conversely, non-replicated data permits local updates but entails the possibility of a single-point failure. The selection of the optimal update approach is contingent upon the particular demands of the system, encompassing variables such as the significance of the data, objectives related to performance, and deliberations regarding fault tolerance.

There exist numerous strategies for updating replicated and fragmented data. This section aims to examine the aforementioned strategies and provide illustrations of their implementation through examples from two prominent vendors (Freisleben et al., 1991).

8.5.1. Eager (Synchronous) Replication

According to Gray et al.'s (1996) research, the process of eager replication ensures complete synchronization of all replicas across all nodes (sites) by modifying all replicas as a single atomic transaction. Implementing eager replication results in a serializable execution, thereby eliminating any potential concurrency anomalies. However, the act of replicating data with enthusiasm can lead to a decrease in update performance and an increase in transaction times. This is due to the additional updates and messages that are included in the transaction. In general, eager replication employs a locking mechanism to identify and manage simultaneous execution (Kemme & Alonso, 2000).

Through prompt replication, data readings from interconnected nodes provide up-to-date information. Reading data at disconnected nodes can result in the retrieval of outdated information. Eager replication systems of a basic nature restrict the ability to make updates in the event of a disconnected node. Eager replication systems facilitate updates among members of a quorum or cluster to ensure high availability. Upon the addition of a node to the quorum, the quorum transmits to the node all replica updates that have been generated since the node's disconnection (Ezéchiel et al., 2020).

8.5.2. Eager Replication and Distributed Reliability Protocols

Distributed reliability protocols (DRPs) exemplify the implementation of eager replication. According to ORACLE (1993), DRPs exhibit synchronous behavior and frequently employ remote procedure calls (RPCs) (Sharma & Kansal, 2013).

DRPs enforce atomicity in transactions through the implementation of atomic commitment protocols, including the two-phase commit (2PC) proposed by Gray in 1979. A two-phase commit protocol (2PC) is deemed necessary in scenarios where a solitary transaction can interact with multiple independent resource managers. This necessity is especially pronounced in a distributed system, as noted by Ozsu and Valduriez (1996). The 2PC protocol facilitates the extension of local atomic commit actions to distributed transactions. This is achieved by mandating that all sites participating in the execution of a distributed transaction reach a consensus to commit the transaction before its effects are made permanent. The concept of one copy equivalence is exemplified by 2PC, whereby the termination of a transaction that updates a logical data item necessitates that the values of all originals of said item are identical.

The opposite of termination is recovery. The issue of restoring a database to a consistent state at a failed site upon its recovery from failure is addressed by distributed recovery protocols. The 2-phase commit protocol also encompasses recovery within its scope (Wiesmann et al., 2000).

8.5.3. Two-Phase Commit (2PC) Protocol

The 2PC protocol operates as follows, as per Date (1999):

The coordinator is a system component responsible for ensuring that all resource managers engage in or roll back their updates in unison. Its primary task is to provide this guarantee even in the event of a system failure occurring during the process.

It can be posited that the transaction has effectively executed its database processing without any errors, thereby resulting in the issuance of a COMMIT operation instead of a ROLLBACK operation at the system level. Upon receipt of the COMMIT request, the Coordinator initiates a two-phase process consisting of the following steps (Reddy & Kitsuregawa, 1998):

1. Initially, the Coordinator instructs all resource managers to prepare themselves to execute or abort the ongoing transaction. In practical terms, it is necessary for every participant involved in the transaction to manually record all log entries pertaining to the local resources utilized, and store them in their respective physical logs. Upon a successful force-write, the resource manager responds to the Co-ordinator with an affirmative message denoted as 'OK.' Conversely, if the force-write is unsuccessful, the resource manager responds with a negative message, 'Not OK.'
2. Once all participants have responded, the Co-ordinator proceeds to generate an entry in its physical log, documenting

its decision pertaining to the transaction. In the event that all responses are affirmative, the determination shall be deemed as COMMIT. Conversely, if any of the responses are negative, the determination shall be classified as ROLLBACK. Subsequently, the Coordinator proceeds to notify every individual of its verdict, or each participant is required to execute a local COMMIT or ROLLBACK operation in accordance with the instructions provided.

In the event of a system or network failure during the entirety of the process, the restart protocol will search for the decision record within the log of the coordinator. In the event of successful discovery, the 2-phase commit (2PC) procedure can resume from its previous state. If the decision is not located, the assumption is made that it pertains to a ROLLBACK, thereby enabling the process to be suitably finalized. In a distributed system, the failure of the coordinator could result in certain participants experiencing delays while awaiting the coordinator's decision. Consequently, during the participant's waiting period, any modifications performed by the transaction through said participant remain locked (Nouali et al., 2005).

8.5.4. Read-Once/Write-All Protocol

The read-once/write-all (ROWA) protocol is an additional replica-control protocol that ensures the enforcement of one-copy serializability. The ROWA protocol is characterized by its simplicity. However, it necessitates the updating of all instances of a given logical data entity prior to the conclusion of a transaction (Parker & Chen, 2004).

The occurrence of a site failure has the potential to impede a transaction, thereby diminishing the accessibility of the database.

Several alternative algorithms have been suggested to decrease the necessity of updating all instances of a logical data item before the completion of a transaction. The relaxation of ROWA is achieved by mapping each write operation to a limited subset of the originals. Quorum-based voting is a widely recognized method in which votes are assigned to copies, and read-and-write operations necessitate the collection of votes and the attainment of a quorum to read/write data.

The three-phase commit protocol is a non-blocking mechanism that effectively mitigates the 2PC blocking issue by eliminating the ambiguity for participants subsequent to casting their votes. The process above is achieved by incorporating a pre-commit phase, which disseminates pertinent details to involved parties, apprising them of an imminent commitment (Cao et al., 2001).

8.5.5. Lazy or Asynchronous Replication

The eager replication update methods are considered synchronous as they necessitate the atomic updating of a certain number of copies. Both the lazy group replication and lazy master replication mechanisms function asynchronously.

In the context of distributed database systems, users who prioritize the ability to perform asynchronous updates may accept a certain level of inconsistency, provided that two conditions are met.

1. Firstly, they will require a precise limitation on the extent of the inconsistency.
2. Secondly, they will demand that the system ensures eventual convergence to established standards of accuracy.

In the absence of said properties, the system undergoes partitioning as the replicas progressively deviate from each other, as noted by Davidson et al. in 1985 (Vardhan et al., 2012) (Figure 8.5).

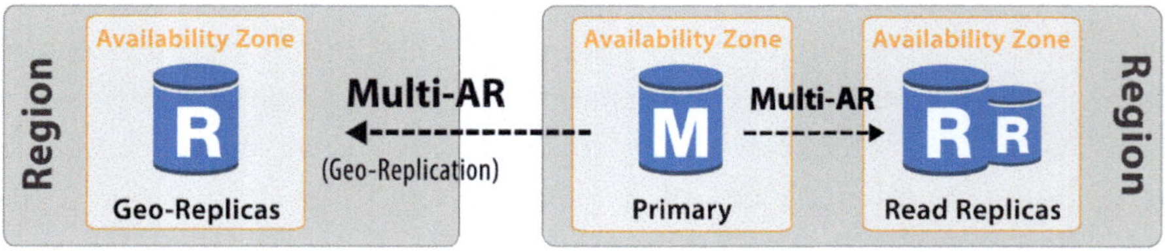

Figure 8.5. *Illustration of synchronous vs asynchronous replication.*

Source: Cloudbasic, Creative Commons License.

8.5.6. Lazy Group Replication

The lazy group replication approach enables unrestricted updating of local data by any node. Upon committing the transaction, a notification is disseminated to all other nodes, directing them to implement the modifications made by the root transaction to the replicas located at the destination node. There exists a scenario wherein two nodes may concurrently modify a shared object and contend to propagate their respective updates to other nodes. According to Gray et al. (1996), it is imperative for the replication mechanism to identify and resolve any conflicts arising from two transactions to prevent the loss of their respective updates (Pacitti et al., 2001).

Timestamps are a prevalent method employed to identify and harmonize indolent-group transactional modifications. The timestamp of the latest update is associated with each object. Every replica update includes the latest value and is labeled with the previous object timestamp. Every individual node is capable of detecting incoming replica changes that have the potential to overwrite previously committed updates. The node performs a comparison between the timestamp of the local replica and the previous timestamp of the update to determine if they are equivalent. If that is the case, then the update can be considered safe. Upon the arrival of a new transaction, the timestamp of the local replica is incremented to match that of the transaction, subsequently resulting in the update of the object value. If the current timestamp of the local replica is incongruent with the prior timestamp observed by the root transaction, the modification in question could pose a risk. In such instances, the node declines the incoming transaction and forwards it for the purpose of reconciliation. The reconciliation process is accountable for applying all pending update transactions in their accurate chronological order.

In contrast to an eager replication system, a lazy group replication system poses the challenge of reconciliation for transactions that are queued for execution. The occurrence of waits is more prevalent in comparison to deadlocks, as the latter requires two waits to transpire (Beloued et al., 2005).

8.5.7. Lazy Master Replication

The primary source of data updates employs a mechanism that defers the dissemination of these updates to the replica nodes until deemed essential. The fundamental concept underlying this methodology is to reduce communication costs by consolidating numerous updates and transmitting them as a solitary entity to the replicas. This action results in a decrease in network traffic, an increase in system scalability, and a performance improvement.

- **The Process of Replication:** Further exploration of the replication process within a lazy master replication scheme is warranted.

1. **Revision at the main node:** Upon updating a data item at the master node, the propagation of the update to all replica nodes is not immediate. Rather, the master node maintains a local log or transaction record to monitor the changes.

2. **Update Propagation:** The process of updating replica nodes is referred to as update propagation. The master node conducts periodic checks on the log or transaction record to determine the appropriate time to propagate the accumulated updates to the replica nodes. The determination of such a decision may rely upon various factors, including but not limited to the quantity of updates, duration of time elapsed, or predetermined thresholds.
3. **Communication and Data Transfer:** These are integral aspects of the system's operation. Upon the master node's determination to disseminate updates, a consolidated update message is transmitted to the replica nodes. The consolidation of multiple updates into a single transmission results in a reduction of communication overhead and subsequently decreases network latency.
4. **Update Application:** Upon reception of the update notification, every replica node proceeds to apply the updates to its respective local data copy. This process guarantees uniformity across the duplicates and synchronizes them with the primary source (Breitbart & Korth, 1997).

- **Advantages of Lazy Master Replication:**

1. Lazy master replication offers a notable *reduction in communication* overhead between the master and replica nodes by consolidating updates and transmitting them as a singular transmission. The aforementioned approach serves to decrease the amount of data transmitted over a network and mitigate the effects of delay, particularly in systems that are dispersed across different geographical locations.
2. *Enhanced system performance* is achieved through reduced communication overhead, resulting in less time spent on propagating updates. The utilization of replicas allows for the master node to prioritize the processing of client requests, while enabling the replicas to synchronize with updates at a self-determined rate.
3. The implementation of lazy master replication can *enhance the scalability* of the system, enabling it to accommodate an increased amount of clients and updates while minimizing performance degradation. The act of batching updates decreases the frequency of communication, which in turn, enables the system to achieve efficient scalability (Daudjee & Salem, 2004).

- **Challenges and Considerations:**

a. The utilization of lazy master replication can result in *improved performance*. However, it can also present *difficulties in ensuring data consistency*. The phenomenon of delayed update propagation entails a situation where replicas may possess momentarily obsolete data. In order to manage conflicts and guarantee eventual consistency, it is imperative to employ methodologies such as conflict resolution, versioning, or timestamp-based ordering.

b. The design of lazy master replication ought to include *fault tolerance considerations*. In the event of a failure of the master node prior to the propagation of updates, replicas may become inconsistent. Various strategies such as replication logs, checks, and ready masters can be utilized to effectively manage system failures and ensure the preservation of data integrity (Malik et al., 2016).

c. The implementation of lazy master replication results in a *reduction of communication overhead.* However, it also introduces an increase in synchronization overhead during the propagation of updates. Coordinating batched updates, delivery tracking, and consistency maintenance among replicas necessitates meticulous design and effective coordination mechanisms for the master node.

The utilization of lazy master replication is a proficient approach in distributed database systems to reduce communication overhead and enhance performance. Through grouping updates and postponing their dissemination, there is a reduction in the amount of data transmitted over the network, an improvement in the system's ability to handle increasing workloads, and optimization of resource allocation. Nevertheless, it poses difficulties concerning uniformity, error resilience, and coordination. Comprehending these factors and executing suitable mechanisms is imperative to guarantee the triumph of idle master replication in dispersed database systems (Pratama, 2011).

Activity 8.1

In a distributed database system, how can the Two-Phase Commit (2PC) protocol ensure atomicity and consistency of transactions involving replicated data across multiple sites?

8.6. SUMMARY

This chapter of the book discusses various aspects of distributed database systems. It covers topics such as fragmentation independence, replication independence, and update strategies for replicated and non-replicated data. The chapter explores real-world scenarios, including a banking system, to illustrate the practical applications of distributed databases. It emphasizes the importance of choosing appropriate update strategies and understanding the protocols, such as the Two-Phase Commit (2PC) protocol, to ensure data consistency and reliability. Distributed database systems offer improved performance and reliability by enabling data storage and access from multiple locations.

Review Questions

1. What is fragmentation independence, and why is it important in distributed database systems?
2. Explain the difference between eager (synchronous) replication and lazy (asynchronous) replication in distributed database systems.
3. What are the advantages and disadvantages of using replication in a distributed database system?
4. How does the Read-Once/Write-All protocol work in maintaining data consistency in replicated data environments?
5. Describe the Two-Phase Commit (2PC) protocol and its role in ensuring transaction atomicity in distributed database systems.
6. What factors should be considered when choosing an appropriate update strategy for replicated and non-replicated data?
7. How does lazy group replication differ from lazy master replication in distributed database systems?
8. Discuss the potential challenges and solutions in achieving data consistency in distributed database systems.

REFERENCES

1. Ahmad, I., Iqbal, S., Jamil, S., & Kamran, M., (2021). A systematic literature review of e-banking frauds: Current scenario and security techniques. *Linguistica Antverpiensia*, *2*(2), 3509–3517.
2. Beloued, A., Gilliot, J. M., Segarra, M. T., & André, F., (2005). Dynamic data replication and consistency in mobile environments. In: *Proceedings of the 2nd International Doctoral Symposium on Middleware* (Vol. 2, pp. 1–5).
3. Bernstein, P. A., & Goodman, N., (1981). Concurrency control in distributed database systems. *ACM Computing Surveys (CSUR)*, *13*(2), 185–221.
4. Bratsberg, S. E., Hvasshovd, S. O., & Torbjørnsen, Ø., (1997). Location and replication independent recovery in a highly available database. In: *Advances in Databases: 15th British National Conference on Databases, BNCOD 15 London, United Kingdom, July 7–9, 1997 Proceedings 15* (Vol. 3, pp. 23–37). Springer Berlin Heidelberg.
5. Breitbart, Y., & Korth, H. F., (1997). Replication and consistency: Being lazy helps sometimes. In: *Proceedings of the Sixteenth ACM SIGACT-SIGMOD-SIGART Symposium on Principles of Database Systems* (Vol. 2, pp. 173–184).
6. Cao, J., Chan, A. T., & Wu, J., (2001). Achieving replication consistency using cooperating mobile agents. In: *Proceedings International Conference on Parallel Processing Workshops* (Vol. 1, pp. 453–458). IEEE.
7. Daudjee, K., & Salem, K., (2004). Lazy database replication with ordering guarantees. In: *Proceedings. 20th International Conference on Data Engineering* (Vol. 1, pp. 424–435). IEEE.
8. Du, W., Elmagarmid, A. K., Kim, W., & Bukhres, O., (1993). Supporting consistent updates in replicated multidatabase systems. *The VLDB Journal*, *2*, 215–241.
9. Ezéchiel, K. K., Ojha, S. K., & Agarwal, R., (2020). A new eager replication approach using a non-blocking protocol over a decentralized P2P architecture. *International Journal of Distributed Systems and Technologies (IJDST)*, *11*(2), 69–100.
10. Foster, E. C., Godbole, S., Foster, E. C., & Godbole, S., (2016). Distributed database systems. *Database Systems: A Pragmatic Approach*, *7*, 361–370.

11. Freisleben, B., Koch, H. H., & Theel, O., (1991). The electoral district strategy for replicated data in distributed systems. In: *Fault-Tolerant Computing Systems: Tests, Diagnosis, Fault Treatment 5th International GI/ITG/GMA Conference Nürnberg, September 25–27, 1991 Proceedings* (Vol. 2, pp. 100–111). Springer Berlin Heidelberg.
12. Grebla, H. A., & Cenan, C., (2005). Distributed database replication-a game theory? In: *Seventh International Symposium on Symbolic and Numeric Algorithms for Scientific Computing (SYNASC'05)* (Vol. 1, pp. 4–7). IEEE.
13. Gupta, S., Yun, H., Xu, H., & Kim, H. W., (2017). An exploratory study on mobile banking adoption in Indian metropolitan and urban areas: A scenario-based experiment. *Information Technology for Development, 23*(1), 127–152.
14. Hellerstein, J. M., Stonebraker, M., & Caccia, R., (1999). Independent, open enterprise data integration. *IEEE Data Eng. Bull., 22*(1), 43–49.
15. Hu, D., Zhao, J. L., Hua, Z., & Wong, M. C., (2012). Network-based modeling and analysis of systemic risk in banking systems. *MIS Quarterly*, 1269–1291.
16. Kemme, B., & Alonso, G., (2000). A new approach to developing and implementing eager database replication protocols. *ACM Transactions on Database Systems (TODS), 25*(3), 333–379.
17. Kumar, N., Bilgaiyan, S., & Sagnika, S., (2013). An overview of transparency in homogeneous distributed database system. *International Journal of Advanced Research in Computer Engineering & Technology (IJARCET), 2*(10), 52, 53.
18. Malik, S. U. R., Khan, S. U., Ewen, S. J., Tziritas, N., Kolodziej, J., Zomaya, A. Y., & Li, H., (2016). Performance analysis of data intensive cloud systems based on data management and replication: A survey. *Distributed and Parallel Databases, 34*, 179–215.
19. Mullin, J. K., (1990). Optimal semijoins for distributed database systems. *IEEE Transactions on Software Engineering, 16*(5), 558–560.
20. Nayak, S., & Chandiramani, J., (2022). A crisis that changed the banking scenario in India: Exploring the role of ethics in business. *Asian Journal of Business Ethics, 5*, 1–26.
21. Nouali, N., Doucet, A., & Drias, H., (2005). A two-phase commit protocol for mobile wireless environment. In: *Proceedings of the 16th Australasian Database Conference* (Vol. 1, 39, pp. 135–143).

22. Ozsu, M. T., & Valduriez, P., (1991). Distributed database systems: Where are we now? *Computer, 24*(8), 68–78.
23. Pacitti, E., Minet, P., & Simon, E., (2001). Replica consistency in lazy master replicated databases. *Distributed and Parallel Databases, 9*, 237–267.
24. Parker, S., & Chen, Z., (2004). Extending invalid-access prevention policy protocols for mobile-client data caching. In: *Proceedings of the 2004 ACM Symposium on Applied Computing* (pp. 1171–1176).
25. Pratama, A. R. I., (2011). Two major issues in data grid replication process. *Jurnal Inovasi dan Kewirausahaan, 7*, 119860.
26. Raja, V., (2012). Global e-banking scenario and challenges in banking system. *Asian Journal of Research in Banking and Finance, 2*(3), 92–101.
27. Reddy, P. K., & Kitsuregawa, M., (1998). Reducing the blocking in two-phase commit protocol employing backup sites. In: *Proceedings; 3rd IFCIS International Conference on Cooperative Information Systems (Cat. No. 98EX122)* (Vol. 2, pp. 406–415). IEEE.
28. Sacco, G. M., (1986). Fragmentation: A technique for efficient query processing. *ACM Transactions on Database Systems (TODS), 11*(2), 113–133.
29. Sharma, A., & Kansal, V., (2013). An evaluation framework of replication protocols in mobile environment. *International Journal of Database Management Systems, 5*(1), 45.
30. Singh, J., (2012). Scenario of e-banking in today's life-a survey. *International Journal of Computing & Business Research, 8*(3), 8563.
31. Sivasubramanian, S., Alonso, G., Pierre, G., & Van, S. M., (2005). GlobeDB: Autonomic data replication for web applications. In: *Proceedings of the 14th International Conference on World Wide Web* (Vol. 2, pp. 33–42).
32. Thomas, G., Thompson, G. R., Chung, C. W., Barkmeyer, E., Carter, F., Templeton, M., & Hartman, B., (1990). Heterogeneous distributed database systems for production use. *ACM Computing Surveys (CSUR), 22*(3), 237–266.
33. Traiger, I. L., Gray, J., Galtieri, C. A., & Lindsay, B. G., (1982). Transactions and consistency in distributed database systems. *ACM Transactions on Database Systems (TODS), 7*(3), 323–342.

34. Ulus, T., & Uysal, M., (2003). Heuristic approach to dynamic data allocation in distributed database systems. *Pakistan Journal of Information and Technology*, *2*(3), 231–239.

35. Vardhan, M., Verma, S., Bhatnagar, P., & Kushwaha, D. S., (2012). Eager computation and lazy propagation of modifications for reducing synchronization overhead in file replication system. In: *2012 Third International Conference on Computer and Communication Technology* (Vol. 1, pp. 331–334). IEEE.

36. Wiesmann, M., Pedone, F., Schiper, A., Kemme, B., & Alonso, G., (2000). Database replication techniques: A three parameter classification. In: *Proceedings 19th IEEE Symposium on Reliable Distributed Systems SRDS-2000* (Vol. 5, pp. 206–215). IEEE.

INDEX

D

E

F

G

H

I